Advance Praise for *Solving Chemistry*

I loved it. The autobiographical tone throughout makes it much more readable than a dry account of the state of chemistry, and of course for me that rang so many bells. You set out so clearly the actual way that science develops by placing it in this personal context.

> Professor Sir David King – former UK Government Chief Scientific Advisor,
> Emeritus Professor of Chemistry at Cambridge

An interesting read. I know what you mean by the completion of Chemistry and I suspect that the same thing could be said of many 'traditional' disciplines. I found the second part of the book really interesting because I was familiar with other elements of the same story and the same problems particularly when it is a matter of the interaction of science/technology with business/academia.

> Lord Oxburgh – former Head of the Department of
> Earth Sciences at Cambridge, past Rector of Imperial College

A real page turner! Love the story. There are so many very good lessons for youngsters (and oldsters!) that can be taken from your experiences. I applaud you for presenting your story so well.

> Professor Harry B. Gray – A. O. Beckman Professor of Chemistry,
> California Institute of Technology

Solving Chemistry is an enthralling account of the interplay between applied and basic chemistry over fifty years. Advances in instruments and measurement enabled dramatic leaps in understanding molecules and reactions. Dr. Bernard Bulkin's intuition led him to the forefront of his field to focus on the most difficult challenges in basic research, in academic leadership, and in industry's development and use of knowledge. This is an inspiring story and invaluable guide to building a career in science aimed at embracing new directions and opportunities and possibly changing the world.

> Professor James Utterback – David McGrath Professor
> Management and Innovation, MIT

Too few memoirs from scientists even try to blend personal experience with history of science. Bulkin combines both ingredients artfully. The personal is communicated with disarming honesty and charm; I often found myself smiling from ear to ear. The historical account is deeply original and provocative. Among its readers will be students of all ages seeking guidance about their careers, because it carries the message that the life of the scientist can be exceptionally fulfilling, can take many turns, and is always a work in progress.

> Professor Robert Socolow, Princeton University

For my teachers, especially Robert P. Bauman,
Walter F. Edgell and Hans H. Günthard

* * *

When we think about the present, we veer wildly between the belief in chance and the evidence in favour of determinism. When we think about the past, however, it seems obvious that everything happened in the way it was intended.

Michel Houellebecq, Atomised

Solving Chemistry

A Scientist's Journey

Bernard J. Bulkin

First published in Great Britain in 2019 by
whitefox publishing Ltd.

www.wearewhitefox.com

ISBN 978-1-91-289206-8

Project management by whitefox
Designed and typeset by Tom Cabot/ketchup
Cover design by Laura deGrasse

Printed and bound in Great Britain by Clays Ltd, Elcograf S.p.A

Contents

An Introduction 9

1. The Discipline: Chemistry 12

2. Being Educated: Learning to Think in the Unnatural
   Way That Is Science 19

3. Becoming a Spectroscopist: How Development of
   Instrumentation and Measurement Techniques Allowed
   Very Complex Problems to Be Studied and Solved 27

4. Perfluorocyclobutane: My First Research Problem and
   an Example of Using Spectroscopy to Solve a Subtle
   Point of Molecular Structure 37

5. Some Thoughts on Molecular Structure: Theory
   and Experiment 49

6. Iron Pentacarbonyl and Amines: Bringing a Whole Bunch
   of Techniques to Bear in Order to Resolve Conflicting
   Views of What Is Happening in a Chemical Reaction,
   and Getting a PhD in the Process 54

7. Some Thoughts on Chemical Reactions 62

8. The Pleasure and Privilege of Doctoral Education . . . 67

9. Far Infrared Absorption of Polar Liquids: An
   Unexpected Observation Needs to Be Explained,
   and This Opens Up the Possibility of Understanding
   More about Liquids 74

10. Vibrational Spectra of Liquid Crystals: Probing of 'Phase Transitions' Becomes Possible, and This Can Be Applied to Biological Membranes 84

11. Some Thoughts on Understanding Condensed Matter . 102

12. Advancing the Technique 105

13. Awards 110

14. Things It Is Important to Know 112

15. Polymers Everywhere: Tackling Ever More Complex Molecules, Their Structure and Reactivity, Becomes Possible 120

16. Crystallisation Kinetics of Polymers: Polyesters; Faster Techniques Allow Us to Watch Crystallisation as It Is Occurring 123

17. Stale Bread and Crispy Cookies: And We Can Apply This to the Complex Problem of Food Spoilage 128

18. Polymer Degradation Processes: Probing Chemical Reactions in Complex Molecules Exposed to Heat, Light and Other Extreme Conditions 135

19. Some Thoughts on Polymer Science 137

20. Transition: A Scientific Life in Academia Exchanged for One in Industry 141

21. The Reproducibility of Results: Laboratory Robotics Change What We Can Measure and How We Can Do It Better Than with People. 149

22. The Measurement of Octane: Using Advanced Computational Methods to Simplify Determination of the Most Important Property of Transport Fuel 158

23. Project Sunrise: The Complexity of Lubrication
and Biodegradability. 167

24. Clean Fuels: Applying Chemistry to the Problem of
Automotive Exhausts, and Using the Science of
Measurement to Optimise Data Collection;
Chemistry Meets Politics. 172

25. Clean Fuels 2: The Thirty Cities Programme,
in Which Our Chemical Knowledge Allows Us to
Change the Competitive Game, and Also to Explain
the Impact to the Public in a Very Simple Way. 185

26. Clean Fuels 3: Beyond Petroleum, Reviving Some
Very Old Ideas and Inventions Becomes Possible. . . . 193

27. Chief Scientist: How University Research at the Most
Prestigious Universities, in Chemistry and Related
Sciences, Shifted to Solve Major Problems, and
This Is a Manifestation of the State of the Discipline. . 198

28. Making Bugs Do Chemistry 215

29. Do You Like Mushrooms? Not Just in the US and
Europe, but Also in China 218

30. Celebration of Mastery for the Discipline
of Chemistry. 229

Acknowledgements 237

Cover image: Rosalind Franklin's 1953 X-ray diffraction photograph of DNA.

An Introduction

WE THINK OF SCIENTIFIC DISCIPLINES AS STRETCHING endlessly into the future. There must always be more to learn, more problems to research, breadth and depth of understanding to fill. That is the way I thought of chemistry when I began to study it at university in the late 1950s, and when I started my first research problem in 1961.

But today I think of my discipline, chemistry, very differently. I believe that the tools that became available to chemists in the 1950s to solve problems of structure, reactivity and properties of materials were so powerful that the vast enterprise of chemistry managed to complete the knowledge of the discipline. Of course, it is not 100 per cent complete: there are always new things to learn. But it is complete in two important senses: there are no more big discoveries to be made, no surprises; and there are no major problems of structure or reactivity that are not extensions of examples that we already understand.

Chemistry is sometimes referred to as the central science, and in one sense it sits in a centre between physics on one side and biology on the other. Neither physics nor biology has the same feeling of completion that chemistry has today. Physics, because the fundamental problems are at the limits of both mathematics and human ability to comprehend, and biology, because the complexity of biological systems is so great. Chemistry always had the advantage that its problems were manageable, could be well defined and then solved, if only the tools were available.

Chemistry is also the central science because having attained the level of understanding that we now have, we can use our knowledge to solve difficult problems in biology, medicine, materials science, environmental science and industrial chemistry, all of which rely on understanding the fundamentals of chemical systems. It is this use of chemistry fundamentals in adjacent fields that is what greater than 80 per cent of the enterprise of chemistry is doing today.

In this book, I am going to present this thesis of the completion of chemistry with some anecdotal evidence, and explain how it happened through personal insights, by describing problems that I worked on myself over a course of about forty years. During this time I too made the transition from working on very fundamental problems of molecular structure and reactions to applications of chemistry.

From my start as a very immature and malleable youth of sixteen, having had a mediocre education in science subjects in high school, I was turned into a scientist by my teachers and mentors. From my family background, I had the work ethic required to become a scientist, and this was, in retrospect, essential, because what I was undertaking was very difficult. But work ethic was not enough. My brain had to be 'rewired'. I learned how to think and observe differently, in what Lewis Wolpert called the 'unnatural nature of science' thinking,[*] because my teachers didn't just teach me facts and formulae but knew what was essential for success in science. Having studied the body of chemical knowledge once as an undergraduate, and learned a lot of mathematics and physics as well, I went through most of it a second time as a graduate student, and learned more chemistry and more mathematics as well. And then when I started to teach as a young assistant professor I learned a lot of it again, until it was deep in my brain. All the while I was practising, using what I had learned to solve problems, and as such 'learning by doing'. Eventually I became someone who could contribute to the project that was chemical science in the second half of the twentieth century.

Chemistry in 1950 was old but immature. As I will show, there were cases where we didn't understand what we were measuring, there were profound disputes about what happened in certain reactions, and because of the large number of poorly understood specifics it was not possible to see the big picture of how structure and reaction fit together. All this changed, and changed rapidly. I want to stress right at the start, as I will do at the end of this book, that this is not an obituary for chemistry; rather it is a celebration of an enormous achievement, one that was unanticipated in science, and is, in my view, still not acknowledged either within the discipline or in the scientific community. It is a great triumph of human mastery.

---

[*] Lewis Wolpert, *The Unnatural Nature of Science*, London: Faber and Faber, 1992.

The Discipline: Chemistry

O F THE SCIENTIFIC DISCIPLINES, CHEMISTRY IS ONE of the oldest. Even putting aside the ancient Greek ideas about earth, air, fire and water, and alchemical experiments in the early centuries of the Common Era, if we think of modern science as dating from the time of Galileo, then it was not long after his death that Boyle, Hooke and others were putting alchemy into its grave and setting the basis for chemistry as a scientific discipline in the century from 1660 to 1760.

Oxford University established a professorship in chemistry in 1683, and Cambridge in 1702. The early occupants of these Chairs established research and teaching laboratories, conducted a range of experimental research and developed a way of teaching chemistry to undergraduates as a part of the natural philosophy curriculum. As early as 1707 it was recommended that all Cambridge undergraduates receive education in chemistry. While some academics objected that chemistry was too dirty a science to be part of Oxford and Cambridge, it met with less objection than biology, because it was not seen as a challenge to Christian religious principles.

Two hundred years later, at the beginning of the twentieth century, while there was widespread acceptance of the ideas of atoms and the elements, and a lot of chemical trends were understood from Mendeleev's periodic table (1869), the idea of atoms being held together as molecules with a fixed spatial arrangement was by no means universally accepted. Moreover, some doubted that the properties of atoms could be maintained if the atoms were linked together to form what we now know as molecules. So eminent a chemist as Wilhelm Ostwald, credited with founding much of the modern science of chemistry and winner of the 1909 Nobel Prize, wrote in 1907 of the postulate of molecules that 'the idea that the elements have disappeared, but are nevertheless present as such, is an indefinite one, too indefinite for scientific use'. In fact, Einstein had offered proof of the existence of molecules, and even showed how to calculate the diameter of a sugar molecule, in his 1905 paper on Brownian motion. Nonetheless, even if there were molecules, chemists of 110 years ago were fairly certain that there would never be any method for 'seeing' a molecule, so it would be impossible to determine whether the atoms were really bound together and in what spatial arrangement. This idea that molecules could never be 'seen' was to be proven wrong, dramatically wrong, in the years to follow.

The controversies over molecules morphed into controversy over the existence of polymers, especially certain biopolymers such as cellulose, and whether they were distinct covalently bound entities or more like aggregations of molecules into colloids. This controversy was carried on by Ostwald's son Wolfgang, and many others well into the twentieth century.

In 1916 G. N. Lewis started to formulate the idea of the role of electrons in binding atoms together, putting in place a systematic approach to what might be the formulae and arrangements of atoms in space. With his approach he was able to show why carbon usually bonded to four atoms, as, for example, in methane, CH_4, and that the C should be at the centre of a tetrahedron of hydrogens (an idea that could already be surmised from work a century earlier on optical activity, i.e. left- and right-handed molecules) while oxygen was more likely to bond to two, as in water, H_2O.

But it was only from about 1920 onwards that real molecular data began to be obtained, with the development of techniques such as X-ray diffraction. Use of X-rays (discovered by Roentgen in 1895) to measure the arrangement of atoms in space was first demonstrated in 1912, but it was applied only to understanding the arrangements of the atoms (ions mainly, that is charged atoms) in simple crystals like sodium chloride (table salt). It took some time until the mathematical methods could be developed to elucidate the structure of a molecule, accomplished definitively for the first case only in 1928. Thus only twenty-five years after most chemists agreed that we could never 'see' molecules, X-ray diffraction allowed us to do just that for the first time. At about the same time several indirect techniques for determining properties at the molecular level were coming into use, such as infrared spectroscopy and the recently discovered Raman Effect, at least in the few laboratories that had the equipment, and the patience, to use them. Nonetheless, despite centuries of work, and the considerable progress of chemistry during the first half of the twentieth century, in 1950 there remained many basic problems that were unsolved, among them problems of:

chemical structure, i.e. how atoms are arranged to form molecules, for example in one of the simplest molecules, water, which we all know is H_2O, meaning it contains two hydrogens and one oxygen, while it was known from the early part of the twentieth century that the hydrogens were attached to the oxygen atom, rather than to each other, so H–O–H and not H–H–O, and even from one measurement that the three atoms are not in a straight line, but rather at an angle,

so not   H—O—H   but

$$\overset{\text{O}}{}$$
H   H

... but what is that angle? And if we think of the hydrogen and oxygen as having nuclei, with protons and neutrons, and surrounding these nuclei there are electrons, then if the chemical bond is due to interactions between the electrons fixing, more or less, the positions of the nuclei, what is the distance between the oxygen and hydrogen nuclei, a number we shall call the bond length?

- **chemical theory**, i.e. how to apply the basic ideas of quantum mechanics formulated by Bohr, Schrödinger, Einstein and others in the early twentieth century to understand chemistry. So, for example, in the water molecule, if the H–O–H angle is 104.5°, does theory tell us why this is so, or could we even have used theory to predict the angle? And given that sulfur falls just below oxygen in the periodic table, and forms an analogous molecule to water, H_2S, why is the HSH bond angle only 92°, and why is the HS bond 40 per cent longer than the HO bond?

- **chemical reactions**, i.e. why do some molecules react and not others, why are some reactions fast and others slow, what does the intermediate stage between reactants and products look like, can we identify and predict the products of complex reactions?

- **electronic and magnetic properties**, i.e. can chemists design and synthesise materials (initially only inorganic, but later also organic/ polymeric) that are conductors, semiconductors, superconductors, or with specific magnetic properties, and that have these properties at temperatures that make them useful for a variety of applications?

- **differences between phases**, i.e. what keeps molecules together (so that they don't turn into gases) in liquids and solids? What is glass and why can it maintain its properties over very long periods of time? Are there glassy states found in biology? Why are there ordered fluids (liquid crystals), and are long-chain molecules, polymers, such as polyethylene or polystyrene, more like liquids, crystalline solids or glasses, or do they form completely distinct phases not found in small molecules?

While much had been accomplished to solve problems of structure, reactivity and phase behaviour from 1920 to 1950, between 1950 and 2000 almost all of the tens of thousands of remaining problems across the range of structures and reactions made possible by the periodic table of the elements were solved. Moreover, chemical knowledge and understanding was so great that it could be applied to creating and understanding complex materials (new plastics, fabrics, television displays, etc.) and biology. Many things that were being done in the chemical industry on the basis of empirical knowledge, that is, a 'try it and see what works' sort of approach, could now be understood on the basis of chemical theory and systematic experimental verification, leading to vast improvements for the industry. Environmental problems that had been caused by the chemical industry could be cleaned up, and future ones could be avoided because there was now a better understanding of the fate of chemicals in air, soil and water.

Chemistry as a discipline had thus reached the state of complete understanding that I am describing (OK, it is not complete but say 97 per cent complete) by the year 2000. What remains is completing the corners of the painting, touching up things here and there, and of course applying all the understanding to other problems.[*]

This is where chemistry is as a discipline today. A glance at any of the weekly or monthly news magazines of chemistry shows that the overwhelming majority of the scientific developments being discussed are about these applications rather than fundamentals of chemistry. For example, the twenty featured speakers at the American Chemical Society national meeting in San Francisco in April 2017 included ten that are applications to biology/genetics/medicine, five that are materials science/engineering, two that are energy or environment. Of the remaining three, one is about industry/academia collaboration, one about imaging of catalysts and one about mass spectrometry, albeit from a scientist who emphasises the utility of her work in biological systems.

That is fine, and the turning of the chemistry enterprise towards undertaking the difficult problems of other disciplines and away from the fundamentals of chemistry is a very productive

---

[*] One might suppose that there are always new chemical reactions to be discovered, but surprisingly there are very few of any significance, i.e. new reactions which are not just another example of a class of reactions that are already well known. In the last forty years, only seven of the chemistry Nobel Prizes have been awarded for new classes of reactions or novel ways of making molecules.

use of the centuries of accumulated chemical knowledge for the benefit of society, but it is not new fundamental chemistry. The possibility of being completed in this way is something that was never contemplated for any scientific discipline, and I believe is largely unrecognised for chemistry today.

I played a very small role in this process of understanding basic chemistry. But the things I worked on are illustrative of what happened between 1950 and 2000. In my first research problem, as an undergraduate, I studied the structure of a molecule, i.e. how the atoms are arranged spatially. Then for my PhD thesis I worked out the details of a particular chemical reaction. As a professor I studied how liquids behaved, what liquid crystals are, how these ideas could be applied to biological membranes and fundamentals of the crystallinity and degradation of large molecules (polymers). Subsequently, as an industrial scientist, I studied how to improve certain basic industrial processes, how to deal with environmental problems, and during much of my career contributed to the development of two of the techniques used to address some of these problems. Multiply what I did by several thousand and you have the collective story of chemistry in the second half of the twentieth century. This is a personal glimpse of that story – the story of solving the totality of problems of chemistry.

Being Educated: Learning to Think in the Unnatural Way That Is Science

I DON'T KNOW IF I WAS SET ON a life in science early on. In high school, and as an undergraduate chemistry student, I was not even sure I was good enough to be a scientist, that is, not just a person who has learned science in a classroom but someone who does original science. Educationally I had the benefit of a quick start. New York City had too few students in its schools when I started, just after the Second World War, so to compensate they reduced the age for beginning primary education, and I entered school at the age of four. Then, by the time I was eleven, the post-war baby boom was ready to start being educated, and there were too many children, so they tested us, and the top 10 per cent based on IQ were told we would do three years, the seventh, eighth and ninth grades, in two. The consequence of these two decisions by the New York City Board of Education was that I, and 10 per cent of my classmates, finished high school at age sixteen.

My high-school grades were good but not excellent, and although I liked science and mathematics courses, and did a lot of them, I also enjoyed English and history. Everyone did biology in the first year at Jamaica High School in the New York City Borough of Queens. Biology as it was taught to us was very boring – it was before a lot of the revolution in molecular biology, but it still didn't have to be so descriptive, and so focused on disease- and nutrition-related study; I suppose that was what the State of New York felt that teachers could actually understand well enough to teach. I did chemistry in the same year, and it was more interesting but I didn't feel I was a natural at it. There was a lot of memorisation and solving of very formula-based problems. Again, looking back from a lot of teaching I did myself, and observation of others teaching at all levels, I think that the quality of that chemistry course was a result of the capability of the man who was teaching us and did not reflect at all the great excitement of chemistry at that time. Then in the next year I did physics, much more interesting, and more difficult. Finally, in the last year, a good course in history of science, which was fun. And mathematics every term for four years. The mathematics did not have to be taught in an interesting way, because it was really just giving us the tools either to function in society for those who would go on to do other things, or as a basis to learn a lot more mathematics for people like me who were going to continue in science or engineering.

By the end of our junior year in high school, we were all thinking about university, and what to study. I thought maybe I would do something like geology or petroleum engineering. My father owned shares in Esso, and I read the news magazines about the company that they sent to him as a shareholder. Their work seemed both interesting and exotic. I told my parents about the direction I was contemplating, and for a few weeks they didn't say anything about it. Then one day they sat down with me at the kitchen table, where all serious meetings in our family were held, and said that unfortunately oil was found in the Middle East, and they doubted that oil companies would ever hire a Jewish scientist or engineer. Hmm, I thought, so no undergraduate education for me in Texas, then. My mind turned to chemistry. I learned later that their view was generally untrue, but we are sometimes limited by our parents and their prejudices. As it happened, the twists and turns that are possible in a career eventually led me, nearly thirty years later, to the oil industry and the problems that interested me in the first place.

Before I could firmly decide on chemistry as an undergraduate major, perhaps as a career, I had a small 'medical' problem to deal with. I had known from the time I was quite young that I did not see colours the same way as others did. For example, everyone spoke about green grass but to me it looked like the colour I called red. Finally, at about age fourteen, I had the standard set of colour blindness tests, where you are asked to see numbers in an array of dots, and I was confirmed as red-green colour blind, a fairly common condition in men. I knew from my high-school chemistry labs that quite a lot of the tests we had to carry out required distinguishing colours, for example in determining whether something was acidic or basic. But I was not sure how much this was a part of the work of a chemist. So I wrote a letter to Professor Arthur C. Cope, Chairman of the Chemistry Department at MIT, stating that I wanted to become a chemist but wondered whether my colour blindness would prevent me from doing this successfully. About ten days later I got a formal reply from Professor Cope. 'Thank you for your letter. Your question has dominated the conversations at the MIT Chemistry Faculty coffee gatherings over the past week. We have concluded that there is absolutely no reason why colour blindness should prevent you from becoming a chemist. It may be that as an undergraduate you will have to carry out some tests requiring distinguishing colours. In that case, ask the person standing next to you in the laboratory for help.' So now I had dealt with both religious and medical problems.

If you are a truly outstanding student in your early teens, it is possible that even in a state school (public school in the US) some teacher might recognise this and urge you on to a career in science or mathematics. But for most of us who become scientists, indeed for most who are successful in science through their lives, grades do not give much direction. Lots of us get good grades, a few get truly excellent grades but may not be inclined to a scientific career, but only rarely is there a prodigy, and that is usually in mathematics. And teachers really have little or no idea of what a scientist does, even less what an engineer does, so have no ability to encourage or discourage a career path. I headed for undergraduate education in chemistry with little guidance but some confidence. The confidence was not born of any knowledge – it came from my growing up on the streets of New York, from a family where you had to argue your point if you wanted to be heard and from a school environment where you had to compete to be noticed.

It turned out I was good enough to be a scientist, though it took me another four or five years to even start to prove it, and I wandered my way, partly by design, largely by accident, around a number of major scientific problems. In contrast to my high-school science teaching, I had the benefit of a mostly excellent undergraduate education at Brooklyn Polytechnic Institute (known then as Brooklyn Poly), in which I didn't just learn a lot of facts, formulae and techniques, but I also learned how to learn new things. I believe it is this more than anything else that distinguishes the mediocre higher education from that which is first rate.

During our four years of undergraduate education we had a lot of laboratory time, during which I was taught how to do experiments with care and precision, while still working quickly. Virtually every course required getting experiments done during an allotted time, and then a lot of homework to write up our observations, do calculations to produce a result from those observations and interpret that result in terms of chemical or physical knowledge. Unless I was able to set up, measure accurately, observe acutely, apply mathematics and theory and interpret the results, I was destined to fail. At first I had no idea what I was doing, but I got better and better at it. And because the days from Monday to Friday were filled with classes, every evening and most of every weekend had to be used for studying, writing up laboratory work and then studying some more.

In parallel with the laboratory work, I was taught a huge quantity of chemistry and physics fundamentals, from organic chemistry to inorganic reactions, structure of molecules as it was understood in the early 1960s, how they reacted, the basics of thermodynamics and kinetics (the rates of reactions) and on to quantum mechanics. We learned the nomenclature of our science, fundamental laws and how to solve problems applying those laws. And again, more and more mathematics, starting not with calculus, but with groups and sets, what underlies algebra and arithmetic really, then on to calculus, where we didn't learn formulae but what the calculus was and where the equations came from, and after several semesters of this, into differential equations. Scientists and engineers took the same maths courses and digested all this mathematics, but the engineers had more of it to apply, while we chemists used just little bits and pieces as undergraduates. We also did three very tough terms of physics to start, and then in our fourth year two more terms of advanced physics – relativity and so on.

The theory that we studied, and the laboratory exercises we were set, were very hard. We were a bright group, but I don't recall that any of us, certainly not very many out of hundreds, found it easy to do well. It was night after night, weekend after weekend, of hard work. If you didn't have the work ethic to persist at this, you dropped out. Many of my classmates dropped out at the end of the first year. We immediately lost touch with them, so focused were we on our internal community. At the end of each academic year, when the final exams for that year were finished, I was so tired that I could barely get myself home. Still, for all of this, I knew somehow that my scientific knowledge was still very superficial. It was, in a sense, the first pass through a range of subjects that, at least for me, would require two or three passes in order to really see the patterns and understand the fundamentals behind those patterns.

I realised only in retrospect that throughout this education a lot more was happening than just stuffing my brain with facts, nomenclature, equations and reaction chemistry, though there was a lot of that. My cognitive processes were being transformed so that I could think like a scientist, and this is not a natural way of thinking. This transformation of an individual's thought processes is why how to do science cannot generally be self-taught, why it is very difficult, and why, if you are trying to create the next generation of scientists, it is important that it be done by faculty who themselves have a deep understanding of how scientists need to think.

This is also true for the education of engineers, whose brains are being reshaped in yet a different way. As a chemist I would go on to pick problems that filled in the basic understanding of structure (arrangements of atoms in space), reactivity of chemicals, the nature of liquids and polymers and reactions of polymers as well. By contrast the chemical engineers would design processes, reactors, optimise the consumption of energy. The education for imaginative design work is very different from that of the scientist.

In addition to the chemistry, maths and physics, we had rigorous training in our first year in how to write. At a reunion fifty years later many of my classmates recalled sitting with Mrs Pollack, our writing teacher, as she tore apart our essays and made us rewrite for clarity of expression, all of us agreeing it was as important to our success in careers as the technical stuff. We also had courses in literature where we were encouraged to argue about ideas, in economics, psychology and the history of art. We were boys with limited cultural experiences, most of us with parents who never went to a symphony, a ballet or an art museum, and needed this as part of our education to develop into proper citizens.

Of course, all this about work is not meant to imply the lack of a rich social life, filled with male companionship mainly (as there was only one woman undergraduate at Poly when I started to study there), while I was a sixteen–twenty-year-old longing for sex of any sort, even a passionate kiss, and not getting a lot of it. As I progressed through the four years, I also took on more and more activities on the campus. By my junior year I was features editor of the newspaper, putting together a full page every week, and in my senior year one of the editors of our yearbook as well. The fraternity I joined had to govern itself, which meant regular meetings on such issues as to whether we remained exclusively Jewish in our membership, who led us, what our dues were, etc., while the newspaper/yearbook activities required lots of independent work as well as teamwork. At some point the university administration decided to have a committee of students and faculty to talk about how to improve undergraduate education, and I was asked to be on that committee. I matured and gained confidence from all of this as much as from my classroom accomplishments. Since I was never going to be a member of a sports team, this confidence laid the foundations for the next stages of my life, inside and outside of science. I am sure that in the best private schools and in many colleges and universities there is an expectation that all students will do some sorts of non-classroom activities that develop these competencies. For me, I could have gone through high school and college doing none of it, but that would have left me with an educational deficit that I would have had to make up later.

Eventually, three-quarters of the way through my undergraduate years, I got to *do* science rather than just *learn* science, and at last found out that I loved the doing of it. Until then, I could have equally well decided to move towards medicine, or something more like engineering. But once I was in a lab doing experiments to solve a problem, my chosen problem, and venturing into new things that I had to master before I could take the next step, I had my niche. At least for the next four decades.

3
Li
Lithium

Becoming a Spectroscopist: How Development of Instrumentation and Measurement Techniques Allowed Very Complex Problems to Be Studied and Solved

MY TIME IN CHEMISTRY RESEARCH WAS MORE or less the forty years from 1960 to 2000, though I continued to do lots of interesting things that used my chemistry during the years from 2000 right up until now. What was the highlight of those forty years? It was the emergence of commercial instruments, readily available, for a number of instrumental techniques:

- in particular what are called spectroscopic methods (the interaction of light with matter) for solving chemical problems,

- the related technique of mass spectroscopy (which is not really spectroscopy at all but the break-up of a molecule into charged fragments, separated according to their mass by passing them through a magnetic field, from which we can deduce what is connected to what),

- separation techniques such as gas chromatography (so that complex mixtures could be separated into their components and each one analysed),

- electron microscopes and other techniques to image at near the molecular level, and subsequently other microscopic techniques such as the scanning tunnelling microscope that actually allow us to image atoms and molecules,

- techniques for the study of structures on surfaces, important to understanding the many reactions in chemical industry that take place between a gas and a solid surface,

- complemented, especially in the period from about 1970 onwards, by relatively inexpensive lasers as light sources, the power of laboratory computers to process the data from the new instruments, and big but readily available central computers to do calculations using approximate methods based on quantum theory.

It was these techniques, coupled with X-ray diffraction, that made it possible to 'see' molecules for the first time, to probe their three-dimensional arrangements, to measure the strength of the bonds between atoms and eventually, once special techniques for ultrafast measurement came into common use, to watch how they were transformed by chemical reactions.

Spectroscopy is an old part of science, going back to early experiments by people like Bunsen who realised that different elements emitted different colours, so that if one had a material, and burned it in a flame, even to the naked eye it was clear that it contained sodium or chlorine or some other element. But when this was combined with a simple instrument containing a prism (and later a construction called a diffraction grating that did the same thing only better), forming a 'spectroscope', the light emitted could be split into its different wavelengths or energies (see box below) and the pattern that resulted was very distinctive. At the time of Bunsen and others this was all about the spectra of atoms, while in what I am about to discuss it is about the spectra of molecules, hence the term used is molecular spectroscopy. Recalling what I said in the Introduction about Ostwald's doubting that molecules existed, and others doubting that we could ever 'see' a molecule anyway, it was the development of the experimental and theoretical basis of molecular spectroscopy that gave chemists the ability to effectively see molecules, not with our own eyes but with our instrumental eyes. It is easy to lose the perspective of how relatively recent this is, because every chemist active today grew up with molecular spectroscopy as part of their education and toolkit. But it is less than a century since people were able to actually believe in this. One of the most well-known books from 1928, Richtmeyer's *Introduction to Modern Physics*, conveys this scepticism in talking about spectroscopy, saying 'In the preceding sections ... we have considered only spectra emitted by atoms. There are other spectra ... which are believed on good evidence to be emitted by molecules containing two or more atoms.'

Now what we learn from visible light, either emitted by matter that is heated, absorbed as light is transmitted through it (which is what makes liquids coloured), reflected or scattered from a surface (one of the causes for the sky being blue), is interesting but only helpful for certain substances. Many have not much interaction

LIGHT WAVES

I refer to light in terms of wavelengths. As you can appreciate from the drawing of a wave shown below,

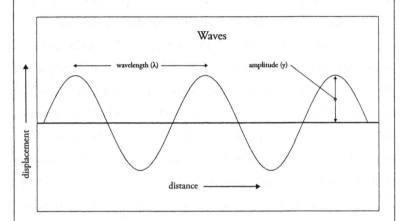

if we think of light as a moving wave, with the spacing between the peaks being the wavelength, then if the frequency with which waves pass by changes, frequency and wavelength have an inverse relationship to one another, that is, as wavelength increases frequency decreases. So if the wavelength has a unit of length such as a centimetre, the frequency has a unit such as waves or cycles/second. If you multiply these together then (frequency) x (wavelength) is in cm/second, which is a speed. As we will see, the frequency of the spectroscopic measurement has a bearing on the ability of the technique to solve the structural problem.

We can use wavelengths of light, usually represented by the lower-case Greek letter *lambda*, λ, energy of the light, E, and frequency of the light waves, usually represented by the lower-case Greek letter *nu*, ν, more or less interchangeably. The equations relating wavelength, frequency and energy of light are very simple, namely: $\nu\lambda = c$. That is, the product of the frequency and the wavelength is, as I said above, a speed, in this case the speed of light, c, and $E = h\nu$, the energy, is the frequency times a constant, h, known as Planck's constant.

Now by light we mean not just visible light, which we can see, but everything from the very high energy of X-rays and Gamma rays to the low energy of radio waves, and in between including ultraviolet radiation, visible light (that is, the energies to which our eyes are sensitive), infrared and microwaves. Before any of the quantum mechanics of absorption and emission of light could make any sense, it was crucial that the foundations were laid in the nineteenth century by Faraday, Maxwell, Herschel and many others that established that all of these energies were light subject to the same physical laws as visible light. Moreover, from Faraday and Maxwell's work, it became clear that this should all be called electromagnetic radiation, having electrical and magnetic properties. Once these properties were established, it was better to represent the wave as

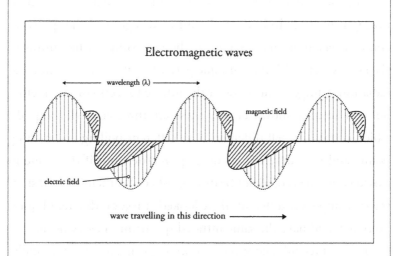

Moreover, and this will be important for some of the later discussions, the electric field is not always in one plane, but it can be forced into one direction or another by passing the light through a material that selects for a particular direction, a polariser.

with visible light, so, for example, you cannot distinguish water from vodka (ethanol) or nail polish remover (acetone) by how they interact with visible light. They all look to our eyes like clear, colourless liquids.* There are shorter wavelengths/higher energies in the ultraviolet region, where a large number of molecules absorb radiation. The great synthetic organic chemist R. B. Woodward at Harvard University realised early in his career that ultraviolet spectroscopy was a key tool for identifying structural components of complex molecules, and devised a set of rules (Woodward's rules) to codify how this identification is made. Nonetheless, even ultraviolet spectroscopy is useful only for distinguishing the structural components of a limited class of molecules.

But there are longer wavelengths, infrared radiation, which do distinguish these. Indeed, in the infrared region of the spectrum almost every molecule, except for the very simplest like nitrogen (N_2) or oxygen (O_2), has a unique pattern of absorption of infrared radiation, a fingerprint. Look at the infrared spectra of water, ethanol and acetone shown below. These are three colourless liquids, which certainly smell different, have different boiling and melting points and very different chemical properties, but if they were in closed containers could not be distinguished. Yet their infrared spectra are completely different. If we looked at 10,000 different liquids no two would have the same infrared spectrum in every detail.

In the three infrared spectra shown opposite, for the three molecules I have been discussing, water, acetone and ethanol, the y axis is

---

* They do smell different, and if we had an understanding of the molecular basis of smell we could tell their structures based on that. Understanding smell at the molecular level is something that may be considered an unsolved problem of chemistry, although it is very much a human senses problem. The 2004 Nobel Prize in Medicine was for an understanding of how our genetic system allows us to distinguish 10,000 smells, but that work did not illuminate the molecular basis for different smells.

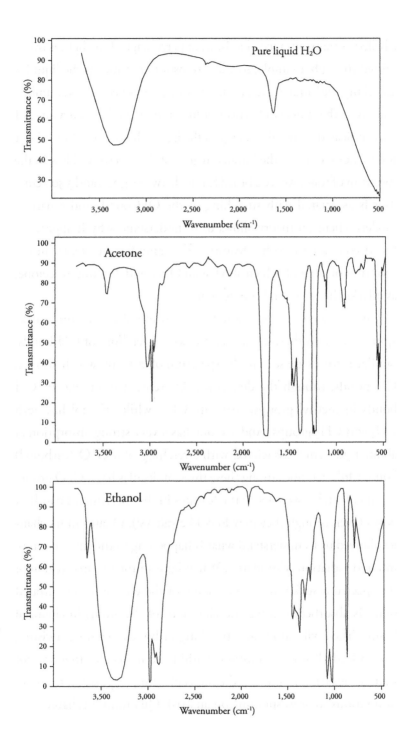

labelled as transmittance, that is, what percentage of the light is transmitted through a sample, so a low transmittance means the liquid is absorbing the infrared radiation. The x axis is labelled as wavenumber, and given the units cm^{-1}. This is a measure of the frequency or, as I have indicated above, the energy of the light. The energies from about 100 to 4,000 cm^{-1} are the infrared region of the spectrum. Now in the spectrum of water we see a broad feature between 3,000 and 3,500 cm^{-1}. This is associated with stretching of the OH bond, and a part of modern chemical theory allows us to understand why it appears so broad compared to other features. The same feature is seen in the spectrum of ethanol, which also has an OH bond, but not in acetone, as it lacks this structural component.

In the acetone spectrum we see some sharp features near 3,000 cm^{-1}, and these are associated with stretching of CH bonds. Similar features are seen in the spectrum of ethanol, which also has CH bonds, though the details are different. Acetone has its CH bonds in methyl groups, structure CH_3, while ethanol has both CH_3 and CH_2 groups. And acetone has a very strong absorption at about 1,700 cm^{-1}, associated with stretching of a C=O (carbonyl) bond. Only acetone among these three molecules has a C=O group.

Incidentally, we refer to the features in the spectrum as bands or peaks (even though they may look like valleys!). Quantum mechanics allowed us to understand what is happening in the interaction of infrared radiation with matter. When light in the infrared region of the spectrum strikes molecules, some of the energy in the infrared beam is absorbed, causing the vibrations of the atoms in chemical bonds. These vibrations are stretching of bonds, bending, twisting and indeed all sorts of motions within the molecules that are not translations (that is, moving the molecule from its position in space) or rotations (that is, spinning on an axis). Quantum mechanics says

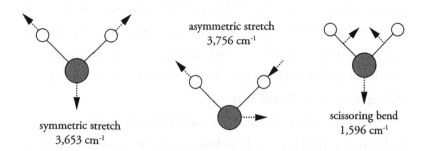

asymmetric stretch
3,756 cm^{-1}

symmetric stretch
3,653 cm^{-1}

scissoring bend
1,596 cm^{-1}

that if the bonds are in a definite environment, say a molecule isolated in the gas phase, the vibrations will occur at well-defined frequencies, they will be quantised. In a liquid there can be a variety of environments which broaden the energies at which the vibrations occur. Each of these so-called modes of vibration occurs at a certain energy or frequency, just as the vibrations of a violin string occur at a certain frequency. I illustrate this above for a water molecule, with the arrows showing how the atoms move.

Now the fact that vibrational modes of molecules could be observed in the infrared region of the spectrum was clear from the early part of the twentieth century, but it took advances in electronics, manufacturing and basic theoretical understanding to be able to both make the acquisition of infrared 'spectra' – the fingerprints – routine, and to be able to interpret the results in terms of things of interest to chemists, like structure of molecules (how the atoms are bound and arranged in space) and function (how the molecules formed of those atoms will react). It had become possible to acquire some of this structural information using X-rays, much shorter wavelengths of radiation, earlier in the century, but this was only useful for crystals, and fairly perfect crystals at that. Most of chemistry takes place in solutions or in the gas phase. The first two easy-to-use commercial instruments for measuring infrared spectra became available from the Perkin-Elmer Corporation in 1952 and 1960, and during the same period from Beckman Corporation. These instruments built on work done earlier in the twentieth century and utilised electronics developed for military applications during the Second World War.

So in the early 1960s not just with infrared spectroscopy but with other techniques such as Raman spectroscopy and nuclear magnetic resonance spectroscopy that I will describe in more detail below, a huge range of previously intractable problems in chemistry could now be solved. And they were solved, so much so that by 1990 virtually every problem of significance in chemistry that existed in 1950 was gone.

Perfluorocyclobutane: My First Research Problem and an Example of Using Spectroscopy to Solve a Subtle Point of Molecular Structure

A T BROOKLYN POLY, STUDENTS IN THREE COURSES, chemistry, physics and metallurgy, were required to do a bachelor's thesis as part of their undergraduate degree, and this occupied a good chunk of time during the final year. Sometime in April 1961 I was awarded a National Science Foundation summer undergraduate research grant that would allow me and nine of my classmates (the top ten!) to start on our thesis during the summer. I gave up the summer job I already had secured working at a camp in Pennsylvania and made the rounds of various faculty members to select an advisor to direct my thesis problem. I decided that I wanted to do something in physical chemistry, that is, the part of chemistry that is closest to physics. Like most chemistry departments at the time, Poly had some strengths and some weaknesses. The most eminent group of chemists there were in polymer science, but we didn't have much contact with them as undergraduates. If we had, I might have chosen to work with Herbert Morawetz or Fred Eirich, both of whom did physical chemistry on polymeric materials. And the greatest physical chemist at Poly was Rudy Marcus, who later won the Nobel Prize, in part for work he did at Poly. But he was on sabbatical leave at the time, so I couldn't meet with him. It is through such coincidences that our scientific lives take turns in one direction or another. In the end I chose to work for Robert P. Bauman, who at the time was still a junior faculty member doing molecular spectroscopy. He had written a book called *Absorption Spectroscopy*, which was about to be published.

The problem that Bauman suggested to me was one of these basic problems of molecular structure. If four carbons are in a ring, and to each carbon are attached two hydrogens, then the resulting molecule, of formula C_4H_8, is called cyclobutane. Now there are two possibilities: that all four carbons lie in a plane, forming a square with four hydrogens above the plane and four below (because that arrangement of the hydrogens on a carbon was already known); or that the ring 'puckers', so that it seems like two triangles bound together. The alternatives look like this:

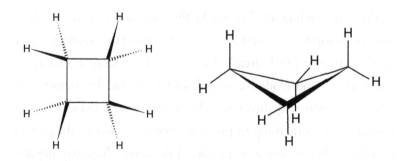

(In both of these diagrams, the carbon atoms are not shown by the letter C, but they form the ring structure. In the diagram of the planar structure shown on the left, the dotted lines indicate that the hydrogen is below the plane of the paper, and the solid wedges mean that the hydrogen is above the plane.)

Now there were three key spectroscopic techniques for resolving this. I have already mentioned infrared spectroscopy, but it does not solve the problem on its own. In the late 1920s, an Indian scientist, C. V. Raman, discovered that when light is scattered from molecules, most of the light has the same energy or frequency as it did when it struck the molecules, but a very small fraction is shifted in frequency, transferring some of its energy into the vibrations of the atoms in the chemical bonds, just as in the infrared region. This is shown schematically below.

Infrared spectra   Raman spectra

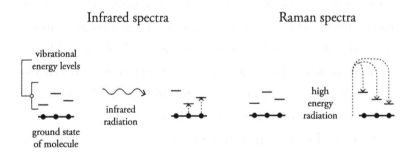

But what is interesting, and useful, is that which energies (frequencies) are absorbed or scattered may differ between the infrared and what came to be called the Raman spectra. And the differences reflect the symmetry of the molecule.

What is meant by symmetry in this context? If we have a triangle, it has what we call threefold symmetry, that is, we can rotate it by 60° and then by 60° again, and then a third time, and after each rotation what we have will be indistinguishable from the starting point. A square, such as the one shown on page 38, will have four-fold symmetry, and we can rotate it through 90° four times, each time having a result indistinguishable from the start. Now if we look at the two molecules above, for the puckered form on the right this fourfold symmetry is broken, because after a 90° rotation what was up is down, and vice versa. Only a 180° rotation will bring us back to the where we started. Another key symmetry element is the existence of planes – we can think of them as mirrors. In the puckered form there are planes through the CH_2 groups opposite one another, so there are two planes of symmetry, one through the CH_2 groups pointing up, and one through the CH_2 groups pointing down. In the planar form on the left, there are many more planes of symmetry. The four carbons lie in a plane, so we can have a mirror in that plane that reflects top and bottom. And there are many mirror planes through the opposite CH_2 groups, and through the centres of opposite CC bonds. So if we think of the problem of the four-membered ring as one of distinguishing between two symmetries, in principle measuring the infrared and Raman spectra of the molecule should solve the problem.

Raman's discovery, in February 1928, was recognised as being momentous, and he received the Nobel Prize in Physics in 1930. It is said that he was so certain that he would receive the Nobel Prize that he booked his ship ticket to travel to Sweden before the prizes were announced. Nonetheless, while a great deal was done to lay the theoretical foundations for Raman spectroscopy over the decades following its discovery, and many spectra were measured, it was only when continuous lasers were invented and the first commercial gas laser, the Helium Neon laser, became commercially available around 1964 that Raman spectroscopy began to be applied to a wide range of structural problems.

The rules for determining how symmetry affects the absorption or scattering of light are called selection rules, and they are derived from a mathematical formulation known as Group Theory. In order to learn how to use this to solve my problem, I had to learn at least some of this mathematics, because, while it had been introduced to us as first-year undergraduates, it is not an area routinely taught in any depth to undergraduate chemists. I learned it mostly in a very practical and mechanical way, just enough really so that I could solve my problem and problems like it. This is how most physical chemists learn group theory to this day. Later, when I was a university faculty member, a colleague, Lou Massa, and I decided to teach a course for our graduate students at City University of New York, where we derived all the equations behind this, and it was only then that I really found out what I was doing. But sometimes you have to concentrate on solving the problem and postpone learning all the mathematics behind it. In any case, as every serious physicist and physical chemist has learned, you need to know a lot of mathematics if you are going to be able to do science, and shying away from learning it because it is too difficult (which it is) just limits what you can accomplish.

I said that there were three techniques for dealing with the structural problem, and so far I have only mentioned two, infrared and Raman spectroscopy. The third is called nuclear magnetic resonance, and when I started working on this problem in 1961 the first commercial instrument, the Varian A60, had just become available, and we had one at Brooklyn Poly. In 1946 two physicists, Bloch and Purcell, had discovered that when molecules are placed in a magnetic field, the energy levels associated with certain of their nuclei split apart, the size of the split depending, as expected, on the magnetic field strength.* But what was unexpected was that the splitting also depended on the chemical environment, very sensitively. Being physicists, they named this effect the Chemical Shift, the dirtiest name they could think of, no doubt, received the Nobel Prize in 1952 and went on to work at other things. But for chemists, this chemical shift became a way to probe and elucidate structure. The difference in energy between the nuclear spin levels was in the radio wave region, very long wavelengths or low energy, so quite different from either infrared or Raman spectra.

If we look at our four-membered ring, and now concentrate not on the carbons but on the hydrogens, we can see that in the planar configuration all of the hydrogens are in exactly the same environment, that is, if you sit on any of the eight hydrogens and look at the others you will see exactly the same thing, but in the puckered form this is not the case. At any given time for a puckered molecule, four of the hydrogens are closer together and four are further apart. So in principle nuclear magnetic resonance, known as NMR, should be able to rapidly determine the structure. (Incidentally, the same technique, renamed MRI for Magnetic Resonance Imaging and configured to produce three-dimensional images rather than lines on a piece of paper, is familiar today as a very sensitive way of probing the body for tumours, spinal dislocations, etc.)

---

* The splitting depends on something called nuclear spin, and this is an interesting aspect of science. No one actually believes that the nuclei are spinning, but, and this is a subtle difference, they are behaving in a way that we associate with spinning particles. We use the term spin because we can make sense of what we are observing by using this word. Likewise, when we talk about light as a wave, and sometimes light as a quantum particle (a photon of light), this works because we understand something about the behaviour of waves in the world we can see, and light is behaving like those waves. For chemists this is sufficient, it allows us to move on and solve our problems. For physicists, questions like the ambiguity between light as particle or wave was the subject of a lot of theoretical development during the twentieth century.

The key difference in symmetry between the planar structure and the puckered one is that in the planar there is a single point, right in the centre of the ring, which is called a centre of symmetry. This is another symmetry element, like the rotation axes and mirror planes. If you start at any atom and draw a line through this point and an equal distance out the other side you will come to the same atom. When a centre of symmetry exists, the selection rules I referred to above say that no absorption/scattering will occur at the same place (frequency, energy) in both the infrared and the Raman spectrum, whereas if that is absent a certain number of peaks should occur in both. Of course, to ascertain this, one needs to measure both spectra very completely and accurately, as there are a large number of peaks possible (more than thirty to be precise).

So what a wonderful scientific problem for me, a student, to have as his first. It had the virtue of being complete in itself, a question to which I could hopefully provide an answer, rather than being part of a larger enterprise as so much of science is, where many students each provide a part of the answer to a very big problem. It required me to learn many new things: techniques for measurement, electronics, mathematics, new chemistry and physics principles. I did this as part of a laboratory with graduate students who became mentors to me, with a faculty advisor who taught me rigour in both experimentation and in how to be very precise in the written expression of the results of my work, in a setting where my peers, other undergraduates, were all working and excited about their own scientific problems.

A research laboratory, as I quickly learned, and as I was to experience for many years, is, at its best, a little family group. In our research group, from my perspective, the parent was Bob Bauman, the older brothers were Stan Abramowitz and Jim Considine, and the older sister was Mary Caulfield. There was also a graduate student named Henry Lee, whom I hardly ever saw. Families are like that. In fact, to my youngest sister Laura, fifteen years my junior, I am the older brother she hardly ever saw. As the eldest of four children at home, here I had the new experience of being the youngest brother, just being initiated into the mysteries of the family. From the day I arrived in the lab, Stan took me under his wing and taught me techniques I needed to know in order to do the experiments to solve the problem.

In those days, the summer of 1961, we did everything for ourselves, and for two reasons: our technique was not in such widespread use yet that commercial accessories needed to make our measurements were readily available, and even where they were available it was thought to be more 'authentic' to make your own, for example the cells that contained the samples. We were working in the infrared region of the spectrum, and glass does not transmit infrared radiation. Instead we used large crystals of rock salt, sodium chloride. To do this, we bought, very cheaply, a big crystal of salt, say 20 cm on a side. What we needed were rectangles about 3 x 2 cm, about 0.5 cm thick. Stan showed me how to take this big crystal and with a razor blade and a hammer cleave off a piece of about the right thickness, and then cleave that into smaller pieces about the right size for the sample cell. But while the salt broke evenly to yield relatively flat surfaces, they were not actually flat, but rather had lots of ridges. I then had to smooth them with various grades of sandpaper, until they were completely flat. And salt dissolves in water, including the moisture on your hand, so everything had to be done with thin plastic gloves on. There was a technique to measure just how flat my surfaces were after polishing, using a sodium vapour lamp. And when I had achieved the desired flatness, I had to be able to get the sample into the cell. For this, I drilled two holes in one of the plates with a little jeweller's drill. Yes, sometimes when I was drilling the holes the plate cracked and I had to start again. After that I wanted the two plates to be a precise distance apart, and glued together so that the material did not leak out. For this I used lead spacers, which I coated with mercury to form an amalgam, and then pressed the plates together. The amalgam formed a seal, and after a few tries I was able to do this well enough so it didn't leak. All this was patiently taught to me by Stan during the first week in the lab.

At the same time, Jim Considine and Bob Bauman worried about cramming enough theory into my head so that I could interpret the data I was going to get. I was given a lot of books to read, and expected to digest them quickly. These included books on group theory, including one by Indian scientists Bhagavantam and Venkatarayudu, of which I still have a first edition, and a little book by the great Caltech pioneer of nuclear magnetic resonance, J. D. Roberts. And what did Mary do during this time, aside from working on her own PhD problem? She taught me not to worry when I didn't get either the experimental or theoretical part right the first time.

In this way I developed equipment to make my measurements, measured spectra, and learned how to interpret them. In the process I was gradually accepted by everyone in the lab as a 'member of the group', which included being someone they could support, help, criticise, tease, even occasionally confide in. Eventually I could also help them, having become the one who knew how to do certain of the lab tasks best.

We did not make all the measurements in our own lab. Other people had some better equipment, and particularly for Raman spectroscopy we were still developing our instrumentation. So I learned that science is a collaborative enterprise, and if you were part of the establishment you could share. At Bell Labs, in New Jersey, George Walrafen had acquired one of the first commercial Raman spectrometers, not yet with a laser but using a 'mercury arc' as light source, built by a company called Cary Instruments. Bob Bauman did not know Walrafen, but at Poly during the year I was working on my thesis problem there was a visiting professor from Bell Labs, John Lundberg (about whom more later). Through him we had an introduction to Walrafen, and immediately an invite to come out and make measurements.

There were also more advanced infrared instruments being developed, especially at Perkin-Elmer Corporation in Connecticut, and we had two contacts there, Abe Savitzky and Bob Hannah. So one day Bob Bauman and I went up to Norwalk, Connecticut and spent a day making measurements. Besides getting some good data I was introduced to the broader community, and people with whom I would be friends for many years.

After all this, eventually, an answer began to emerge. First of all, we did not study cyclobutane itself, but its analogue, where all the hydrogens are replaced by fluorine atoms. Why? Several reasons, but mostly because perfluorocyclobutane, as it is called, can be kept as a liquid at room temperature and so is easier to handle. Through contacts at DuPont, who were experimenting with it as a propellant to blow whipped cream out of cans, we were able to get a cylinder of the stuff of very high purity. This molecule had been studied before, with one paper in 1950 concluding that it was puckered, another in the same year that it was planar, a paper in 1952 favouring the puckered structure, and one in 1955 claiming clear evidence for planarity. Science can be so definitive! Our results appeared in the July 1966 issue of the *Journal of Chemical Physics* and brought all these data together, as well as our new data and interpretations, to show why the puckered structure was the correct one. Most of this was about careful, more extensive measurements with a sample of higher purity than anyone had previously used. We also had at our disposal a more extensive base of theoretical principles that helped in analysing the data than was available to some of the previous investigators.

There was one interesting thing that did not really fall into any of these categories. I have said that in the nuclear magnetic resonance spectrum the planar structure should show a single peak while the puckered structure should distinguish between the different environments of the fluorine or hydrogen atoms. Well, the 1955 paper I referred to above, authored by a pioneer in the techniques, Jack Shoolery, was an NMR measurement that showed this single peak, hence concluding planarity. We knew about this, but thought that what was happening was that the molecule was flipping from one puckered form to the other, as in this diagram:

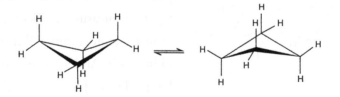

so that on average it appeared planar. This happens for the NMR measurement because it is measured at very low frequencies, hence sees an average, whereas the infrared spectrum is at much higher frequencies. Shoolery's work was very early in the development of the technique, and he did not appreciate this averaging problem. It is like comparing a camera with a very fast shutter speed with one that is much slower. Indeed, a rough calculation is that the infrared measurements look at the molecule in less than a millionth of a millionth of a second, while the NMR measurements were 10,000 times slower, so the molecule could flip 10,000 times during the course of the measurement. Reasoning that temperature should slow this down, we dissolved our molecule in a solvent that froze only at very low temperatures and measured the spectrum as we lowered the temperature. Sure enough, at the very lowest temperatures we could achieve back then, we began to see the band split into its components. Once again, using an approach from another area of physics, Bob Bauman taught me how to use the measurements to calculate how high the energy barrier was to the flipping.

But then we failed to capitalise on our work. Because there was a well-known molecular flipping that we all learned about as undergraduates, not for the four-membered ring we were working on, but for a six-membered ring, cyclohexane.

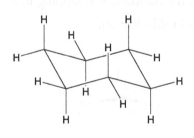

Sure enough, its NMR spectrum also showed a single peak, and we talked about trying to measure its spectrum at lower temperatures. But we didn't do it. And a few months later, Fred Jensen of the University of California at Berkeley published the results of this experiment. That paper turned into a much-cited landmark result, whereas our paper was just a little piece of the overall puzzle of chemistry.

5

B

Boron

Some Thoughts on Molecular Structure: Theory and Experiment

THE ORIGIN OF THE FIRST KEY IDEA in molecular structure, the tetrahedral arrangement of atoms around a central carbon, really was in the need to explain 'optical activity' in the nineteenth century. Optical activity manifests itself by passing light through a material that polarises the light, and seeing how the material rotates the polarisation. For most synthetic chemicals in solution no such interaction with light occurs, but it had been observed in the nineteenth century that quartz crystals interact with polarised light in this way. Subsequently, Pasteur observed this phenomenon for naturally occurring sugars in solution.

While Pasteur was able to pick out left- and right-handed crystals of tartaric acid, there was a puzzle as to how even in solution certain substances could maintain right- and left-handedness. If

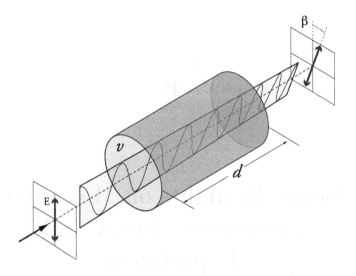

Incoming light is passed through a polariser, and then the polarisation is rotated by the optically active material.

you approach this without much reflection, you would think that as molecules tumble around randomly they would not maintain rotation in the same direction. But of course a left-handed screw is still left-handed no matter its orientation in space, and if the light interacts with handedness, then it doesn't matter what the orientation is. But what would give rise to handedness in the arrangement of atoms? In 1874, van't Hoff and LeBel independently proposed that this phenomenon of optical activity in carbon compounds could be explained by assuming that the four chemical bonds between carbon atoms and their neighbours are directed towards the corners of a regular tetrahedron. If the four neighbours are all different, then there are two possible arrangements of the atoms around the tetrahedron, which will be mirror images of each other. This led to the first real understanding of the three-dimensional nature of molecules.

In the early twentieth century, the ideas of G. N. Lewis, Linus Pauling and others gave quite a lot of insight into how atoms, particularly in organic molecules with carbon backbones, incorporating hydrogen, oxygen and nitrogen, formed chemical bonds, and how these bonds would be arranged in space. However, the details of bond angles, bond lengths, and such things as right- and left-handed molecules needed instrumental techniques for unambiguous determination and measurement, and not many such measurements were made before 1950, especially in gaseous and liquid phases.

The Lewis approach was not to prove so useful for two major areas of chemistry.

Firstly, the structure of molecules containing the so-called transition metals, the middle of the periodic table, where you find elements like iron, nickel, molybdenum and many more, where the larger number of electrons present (and some details which I can't explain here about how those electrons behave in the transition metals and other heavier elements) lead to a much wider variety of structures than we find with organic molecules.

Secondly, the interaction of the molecules, both organic and inorganic, from the simplest (like O_2 or N_2) to the most complex, with light of all energies, but starting with visible and ultraviolet

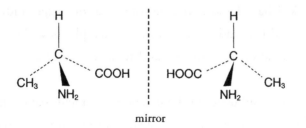

How left- and right-handed structures arise.

light, energies which we knew were required for the electrons in the molecules to be 'excited' to higher energy states.

To explain this required a completely new theory of bonding, based around quantum mechanics, which came to be known as molecular orbital theory. This was originated in the early 1930s by Robert Mulliken and others, who soon found themselves in conflict with Linus Pauling. After the war, Mulliken at the University of Chicago, Coulson in Oxford, Pople at Cambridge and many others progressed this theory and started to apply it to complex molecules and ions, predicting structure and spectra with greater and greater accuracy. At the same time, more and more measurements became routinely possible to confirm the predictions, and in some cases to challenge theory to do better.

As sometimes happens in science, a well-timed, clearly written book can change things. In 1960 a young PhD named Harry Gray went to spend a postdoctoral year with a great Danish scientist, C. J. Ballhausen, and one product of their collaboration was a book called *Molecular Orbital Theory*, which appeared in 1964, by which time Harry Gray was on the faculty of Columbia University in New York. Their explanations made the molecular orbital approach accessible to large numbers of chemistry faculty and students for the first time. In my copy, which I used in the mid-1960s to teach students in the chemistry PhD programme at City University of New York, I have handwritten notes on various pages indicating that I wanted my students to show, for example, how the authors got from equation 1–10 to 1–11 and then to 1–12 as a homework problem.

It was this confluence of theory and experiment that opened up the widest range of structural problems, not just organic molecules but across the range of molecules in the entire periodic table, for solution. So, for example, we could now understand why an ion like $PtCl_4^=$ should have a square planar structure, whereas MnO_4^- should have the Mn atom at the centre of a tetrahedron ...

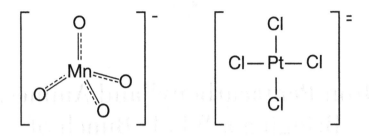

... and thousands of similar problems. This build-up of experimental structure determination with theoretical foundation was reflected in the teaching of inorganic chemistry as well, which went through a major change in the 1960s. When I took my undergraduate course in inorganic chemistry in 1960–1, using a recently published text, *Inorganic Reactions and Structure*, by Edwin Gould, we focused on lots and lots of reactions of the various elements. By the time I was teaching this material in 1967 the standard text was *Advanced Inorganic Chemistry* by F. A. Cotton and Geoffrey Wilkinson, which built on Ballhausen and Gray to give the theoretical and experimental basis of the subject.

6

C

Carbon

Iron Pentacarbonyl and Amines: Bringing a Whole Bunch of Techniques to Bear in Order to Resolve Conflicting Views of What Is Happening in a Chemical Reaction, and Getting a PhD in the Process

I WENT TO PURDUE UNIVERSITY AS A GRADUATE student to work with Professor Walter Edgell. He was a well-known spectroscopist; I had met him once at an American Chemical Society meeting and liked him. Bob Bauman had been an undergraduate at Purdue, and while waiting for his wife to graduate he had done a master's degree with Edgell. Bob Hannah, who I had met at Perkin-Elmer, had done his PhD with Edgell. I wanted to be far from New York City, in a campus setting after four years commuting to university, and based on my grades Purdue was the best university I could get a place in. I had an offer from Minnesota, probably equivalent in reputation to Purdue, and I had met John Overend, who was one of the people there doing work of interest to me. Somehow I didn't take to him, and on that basis, and also that Minnesota was much further from home, and much colder, I chose Purdue. In fact, at Minnesota the better-known person in my field was Professor Bryce Crawford, and if I had met him I would probably have gone there. It was not to be, so at the end of the summer of 1962 I packed up my belongings and headed to Indiana.

I loved my time at Purdue, the several communities there of which I became a part, and the lab in which I worked. But it limited me, or I limited myself to be more honest. I believe that I had the ability to get grades as an undergraduate that would have got me a place in the doctoral programme at Harvard, Berkeley, MIT or Caltech, one of the top universities in chemistry at that time and to this day. But during my first two years as an undergraduate I just didn't know how to study and concentrate on my work, and my grades were mediocre. The third year was better, and the fourth year outstanding, but the damage was done. This limited me later because I wanted an academic post, and all the best ones went to the Berkeley, Harvard and Caltech PhDs. Well, we get what we earn, mostly. And try to make the most of it.

Now I became a member of a new family, the Edgell group, and a big community that was the chemistry department at Purdue. At the time Purdue chemistry comprised about fifty faculty members, 400 graduate students and numerous other postdoctoral fellows and support staff. Edgell's group was about eight when I started, and soon grew to about twelve. It included a postdoctoral fellow, Bob Bayer, Mike Dunkle, who was finishing up his PhD degree, Sara Ann Millard and John Cengel, graduate students whose theses were already well underway, and some others. Also in my class entering in 1962 was Bill Risen, who had come from Georgetown University to work with Edgell, and another graduate student from Taiwan, Shin-Shin Yang, chose to join this group as well. I have talked about research groups as families, and this was especially true for me during my time at Purdue. Doctoral students refer to their thesis advisor as a scientific parent. Walter Edgell and I became very close; most of his group had lunch with him several days a week, where we talked about all sorts of things, from science to politics to religion. I joined him at his home for Thanksgiving dinner on several occasions, and we took many road trips together to other labs, conferences, etc. In many ways I felt closer to Walter than to my own father and our relationship continued after I graduated. When I was a young professor in New York, Walter's wife was very ill with a rare form of cancer and was being treated at the National Institutes of Health in Maryland. He would travel to visit her, and then come up to see me in New York, needing some time to just take his mind off what was happening. We generally started with dinner, and then found various amusements that kept us busy through the night, finishing with breakfast at about 5 a.m. before I sent him off to a hotel and stumbled into bed myself.

Edgell's lab had evolved over time, so that when I got there I found that he was less interested in the basic problems and instrumentation development of spectroscopy, where he had made his reputation, and more in certain fundamental problems of structure and bonding in a class of molecules called metal carbonyls, formed by the reaction of metals such as iron, cobalt and nickel with carbon monoxide. Now these problems did not really interest me all that much, although they should have, because understanding the fundamental nature of the chemical bond between atoms was one of the frontier areas of chemistry at the time. However, there was a scientific problem that was a side interest of Edgell's that did intrigue me, and formed the basis of my doctoral thesis. It came about as follows.

In studying structure and bonding, Edgell wanted to measure and interpret the infrared spectra of three species, of formulae $Ni(CO)_4$, $Co(CO)_4^-$ and $Fe(CO)_4^=$. These three, one molecule and two ions (charged species, as shown one having a single negative charge and the other two extra electrons), actually have the same structure and number of electrons, but they have increasing negative charge as you go through the series, and Edgell's hypothesis was that the infrared spectra could tell how the bonds, in particular the carbon–oxygen bond, and possibly the metal–carbon bond, 'coped' with this charge. Did the extra negative charge strengthen or weaken the carbon–oxygen bond, how much, and why? It was a clever, well-designed experiment. Nickel carbonyl had been known for decades, and though highly toxic was easily obtained. But how to make the other two ions? As it happened a team led by Irving Wender, a well-known chemist, and including Gus Friedel, a very good infrared spectroscopist, had published a paper in 1955 saying that when iron pentacarbonyl, $Fe(CO)_5$, again a readily available, albeit highly toxic and flammable liquid, was dissolved in certain nitrogen-containing compounds called amines, it decomposed into two species, $Fe(CO)_6^{++}$ and $Fe(CO)_4^=$. This was ideal, a simple reaction, and with the added bonus of being able to study the positively charged ion as well. This had been challenged in 1958 by W. Hieber, the world's greatest expert on metal carbonyl chemistry, in Munich, who said that no such thing happened, indeed there was not really a reaction at all, merely a loose association between an unchanged $Fe(CO)_5$ and the amine solvent. A student in Edgell's lab, who had already graduated before I arrived, Meiling Yang, had carried out this reaction trying to make $Fe(CO)_4^=$, but the resulting products seemed nothing like what was expected from either Wender's or Hieber's work. As it turned out, her experimental work had not realised the importance of eliminating all water from the amine liquid, and the reactions she observed were of water with $Fe(CO)_5$. Subsequently, Robert Bayer proved that this was what was happening in Meiling's experiments, but he left to take up a faculty position at Duquesne University before he could figure out what was really going on. In the meantime, Edgell found another route to $Fe(CO)_4^=$ and moved on. But the question of whether Wender's or Hieber's conclusion was correct or not was unanswered. I decided to work on this problem.

This turned out to be a very good PhD thesis. For one thing, it had, like my undergraduate problem, a history, so it was possible to resolve something, there were several techniques that could be applied to the problem, and there was a definite outcome – what actually happened in the reaction – assuming I could figure it out. This is much better than many PhD problems that students take on, where they have to wander around an area for a while to see if there is something interesting to learn. There was another benefit to this problem, and I only appreciated this later: to get the correct scientific answer required the very best technique of measurement, because what happened in the reaction mixture was very sensitive to water, as I knew, and to even traces of oxygen (which I found out quickly), and by very sensitive I mean that less than one part water or oxygen in a million parts of amine would be acceptable. If this wasn't bad enough, I also found that some of the reaction species were sensitive to visible light, and so many of the experiments had to be carried out in the dark. In a later phase of my work, I had to carry out a series of measurements of conductivity in solutions of different concentrations, where the crucial thing was that each sample was subjected to identical treatment, from the time the liquids were combined until the measurement was finished. If the first sample took, say, one minute from start to finish, the following 100 measurements had to be done between 59.5 and 60.5 seconds for the data to be valid. These exacting experimental conditions required development of techniques and skills which served me well throughout my scientific career, not just in measurements I made, but in finding errors in those of my students and others.

One evening after my thesis was more or less complete Edgell gave a talk at one of the chemistry department seminars in which he described the journey which culminated in my work. One of the other faculty members, Grant Urry, was extremely nasty about it, not in public, of course, but in comments to his own students that were repeated back to me (one of whom, Merle Millard, was married to Sara in our group) and others, being scornful of Edgell and his students (not me, but my predecessors) for not appreciating the sensitivity to moisture and oxygen. As if he could never have made such mistakes himself! In the course of my graduate career, and later in my own research career, I saw many examples of this sort of scornful attitude on the part of scientists.

Well, my entire PhD thesis, which was submitted in June 1966, consisted of about 160 pages, and there were another forty pages of the same findings in the format for publication. Indeed, one paper from this thesis had already appeared in the *Journal of the American Chemical Society* even before I graduated. But in fact the whole of the 200 pages could be summed up in three reactions, namely, here is what Wender and friends said, here is what Hieber claimed, and this is the right answer. After that the following 199½ pages are: now here are the data to prove it. Here are the three reactions where the amine is represented by >NH:

$$> NH + Fe(CO)_5 \rightarrow Fe(CO)_6^{++} + Fe(CO)_4^{=} \ldots \ldots \ldots \text{(Wender, 1955)}$$

$$> NH + Fe(CO)_5 \rightarrow >NH \cdot Fe(CO)_5 \ldots \ldots \ldots \ldots \text{(Hieber, 1958)}$$

$$> NH + Fe(CO)_5 \rightarrow >NC(O)Fe(CO)_4^{-} + >NH_2^{+} \ldots \text{(Bulkin \& Edgell, 1965}$$

$$\rightarrow >N(H) \cdot Fe(CO)_4^{+} + >NC(O)H + >NH$$

One way of looking at these three reactions is in the context of the rapidly changing ability to understand what was happening in a reaction vessel when two liquids were mixed. Imagine that our ability to do this was so poorly developed that highly reputable scientists could come to such different answers. But that was the case, and this can be multiplied by thousands of similar reactions, particularly in inorganic and organometallic chemistry, in the early 1960s.

So everything in this PhD thesis made a nice coherent story, showed what was wrong with previous proposals and gave a new answer which seems to have stood the test of time. But we missed one thing. The reaction we had studied, and correctly explained, was one of a class of similar reactions. There were already several similar ones in the literature. But we didn't see this, we were spectroscopists really and not reaction chemists. Fred Basolo and Ralph Pearson at Northwestern University were the latter, and just six months after I finished my PhD I heard a talk by Pearson in which he was able to generalise all these reactions as to what they were. So science is done – you gradually explain all the pieces, and then someone with insight sees how these pieces fit together as an intelligent whole, and with that another part of chemical sciences is put into the 'explained' category. We were part of explaining the pieces, and at least we did that correctly; as long as incorrect explanations were extant in the literature, my reaction would not have added weight to the generalisation. And this is an example of another part of the beautiful painting that is chemistry today. Yes, there are a lot of possible reactions, but gradually we have seen that they fall into classes, so that we can often predict that on mixing A and B there will be a certain set of chemical transformations, because even if this has never been done before we have seen hundreds of other examples just like it.

Some Thoughts on Chemical Reactions

I T IS USEFUL TO THINK ABOUT CHEMICAL reactions from a few different perspectives. We mix A and B and get C and D. Maybe a bit of heat has to be applied, maybe a lot of heat is given off. Perhaps there is an explosion. These are the simple kind of reactions. Then there is the reaction either where electricity is put in, or it is possible to get electrical energy out, for example the reactions that occur in batteries. Well, electrical energy in or out means electrons are added or taken out of the system. Chemistry is about what happens to electrons, chemical bonds are formed from electrons in different atoms, but there is a huge class of reactions that are electron transfer reactions. These reactions have been known for a long time, but there was very little understanding of how they occurred until pioneering work on two fronts, by Henry Taube for complex ions, and by Rudolph A. Marcus doing basic theory. Both these efforts received the Nobel Prize, in 1983 and 1993 respectively. It is interesting that Taube's effort started when he was asked to teach a course in Inorganic Chemistry to graduate students at the University of

OXIDATION/REDUCTION

Chemists have traditionally thought in terms of oxidation states, that is, a simple element, say sodium, symbol Na, as an element is said to have an oxidation state of zero, meaning that it has the same number of protons in the nucleus, with positive charge, as electrons, with negative charge, hence a net of zero. But if the Na transfers an electron to a chlorine atom, Cl, then the sodium is written as Na^+, indicating that it has one more proton in the nucleus than it has electrons, and the Cl is written as Cl^-, since it has an extra electron. The sodium oxidation state is now said to be +1, and the chlorine oxidation state is -1. Another way of speaking of what has happened in this electron transfer reaction is that the sodium has been oxidised, and the chlorine has been reduced. Hence electron transfer reactions are also called oxidation-reduction reactions.

Chicago and couldn't find a textbook that dealt with the mechanisms by which the reactions happened, only descriptive chemistry. It eventually led to a classic paper he published in 1952, about which he later lamented that it had to be done in terms of valence bond theory, because there was not yet sufficient molecular orbital theory to apply (as discussed in my earlier reflections on bonding).

The organic chemists accounted for a huge amount of the development of synthetic methods during the twentieth century. This effort was driven, in part, by a whole strand of chemical research, attracting some of the greatest scientific minds, to perform the total synthesis, from readily available starting materials, of very complex molecules such as quinine, chlorophyll, cholesterol, Vitamin B12 and many others. Of course, the first problem was to figure out, using the spectroscopic techniques described earlier, exactly what the structures of these molecules were. Then many clever steps were required to go from starting materials to final products. It was through these syntheses (most of which were not commercially advantageous over the natural products extracted from plants or other living systems) that a large number of important new reactions were developed. Perhaps as important, the organic chemists made observations about why some reactions were fast and others slow, identified large families of reactions and developed methods using other elements than carbon, oxygen, hydrogen, nitrogen and sulfur, elements such as boron and phosphorous, to produce complex molecules. And ultimately, in the 1960s, they brought molecular orbital ideas to bear in order to understand the so-called stereochemistry of reactions, that is, how to make a very particular arrangement of atoms in a molecule, of which the most important subset is how to make left- or right-handed molecules.

Another important aspect of understanding chemical reactions was trying to elucidate the pathway from reactants to products. We can look at it this way. If the reactants are stable molecules, and the products are also stable molecules, then there must be some sort of energy 'barrier' between reactants and products. Otherwise they would be converting back and forth with each other all the time (and some do, which we call a system at equilibrium). So we require energy to get over this barrier. A first question was how the energy gets into the reactant molecules and stays there for long enough so that the molecule can begin its transformation. How this happens was hypothesised by Frederick Lindemann in 1922 and developed as a theory of reaction energetics by several others over the next fifty years. At the top of the barrier, or near the top, we have something that is neither the reactants nor the products, but something in between. Now there are broadly two possibilities, namely, that the in-between state actually exists as a distinct species, in which case it might be possible to

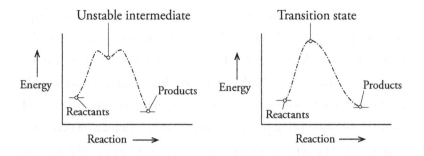

make measurements on it, even find a way to trap it, so it is called an unstable intermediate, or, the in-between state is just a fleeting place on the energy scale that the atoms/molecules pass through, in which case we call it a transition state.

A lot of the effort during the twentieth century, and very successful it was too, was devoted to elucidating these reaction pathways, understanding when intermediates existed, and what the nature of a transition state was. All of this required the most sophisticated use of the spectroscopic techniques I have mentioned, and a lot of theory. As we progressed into the last quarter of the twentieth century, it became possible, because of lasers, arrays of fast detectors and powerful laboratory computers, to make measurements on shorter and shorter time scales, eventually down to a millionth of a millionth of a second, to address these fundamental issues. In Germany, Professor Manfred Eigen showed how these fast detectors could be exploited by subjecting a chemical system at equilibrium to a temperature pulse that would displace it, and then watching how it relaxed back. This further opened up our knowledge of the rates and mechanisms of reactions.

So I witnessed a progression of understanding of chemical reactions during the second half of the twentieth century that was no less dramatic than that in the elucidation of structure. Indeed, the flow of chemical science understanding began with the development of powerful tools for solving the structure of complex molecules, in three dimensions. Then the development of synthetic methods enabled one to build these complex structures from simple molecules. With those synthetic reactions mastered, chemists could see that common principles led to classes of reactions, and that once you see these patterns it is easier to predict the products of a new set of reactants. Applying the same techniques to inorganic and organometallic reactions as had been done with organic molecules extended the reach of reaction chemistry to most of the periodic table. And in parallel there was developed a theoretical framework for the pathways that reactions take that allows one to understand rates of reaction at a fundamental level, using quantum mechanics as it evolved to explain molecular bonding and energies. Building on a vast literature of organic and inorganic reaction chemistry developed over centuries, and with the widespread availability of powerful technological tools, all of this systematisation and fundamental understanding was accomplished in less than fifty years of chemical research from 1950 to 2000.

The Pleasure and Privilege of Doctoral Education

I HAVE MENTIONED SOMETHING ABOUT THE WONDERFUL communities of which I was a part at Purdue. I look back, even after more than fifty years, on my time as a graduate student as one of great pleasure, and yes, privilege. There are, or there were then, really three parts to my doctoral programme. The first is pretty tedious, but necessary. I had to take courses for two years, to build my knowledge beyond undergraduate education, across a range of chemistry, physics and mathematics. Part of this was to accomplish, as I mentioned earlier, the need for two or three passes through a subject to really understand it in depth. This was the second pass. I have seen doctoral programmes, particularly in the UK, that have very little graduate coursework, and I think this disadvantages the graduates in terms of their ability to do research – and this will become clearer as I describe some of the things I did during my research career which required the mathematics, for example, that I learned as a graduate student, and certainly some of the more arcane aspects of quantum mechanics, statistical mechanics and thermodynamics. You can learn these things on your own, if you have to, but it is a lot easier with a teacher, a textbook and the discipline a course imposes.

Another part of this first phase of the doctoral programme was certain examinations. The first set were called qualifying exams, a characteristic then and still of most chemistry PhD programmes in the US, though the format varies. In my case it involved three written examinations of four hours' duration each, covering my choice of any three out of the five fields of chemistry: physical, organic, inorganic, analytical or biochemistry. These required a comprehensive knowledge of everything I should have learned as an undergraduate as well as the first year of graduate school. I had to pass all three of these exams at one sitting, and if I failed even one I could repeat all of them once. At Purdue, these exams were a 'weeding-out' process, because the chemistry department accepted students with doubtful backgrounds or abilities to succeed. Why do this? Because graduate students were needed to teach the laboratories and support faculty for a huge number of basic undergraduate courses. There were something like 105 graduate students admitted to chemistry in 1962, and I think that thirty-three of us passed the qualifying exams.

Of those who did pass, there were more courses in the second year, and maybe half of the third year even. Sometime at the end of the summer after two years, we took another set of exams called preliminary exams. This had three parts. A four-hour written exam on thermodynamics, advanced quantum and statistical mechanics, for those of us who were physical chemistry students, was the first part. If you passed that, then you were asked to write an original research proposal that was not related to the specific area of research in which your thesis would be done. Assuming this written submission was judged meritorious, then your thesis committee (advisor plus three other faculty) would conduct an oral examination cover-

TESTING STUDENTS

Given that I went on to an academic career, in which I composed, administered, graded and relied on thousands of examinations both written and oral, I have had occasion to reflect on what we are doing. Is testing a crucial part of teaching? Do the tests we give select for what we are trying to establish? In the case of the graduate-school examinations I have just discussed, the objective is presumably to determine who among the graduate-school intake is capable of the original thought required to attain a doctoral degree. But there is absolutely no evidence, not a single experiment that says that those who failed these exams are less able to do this than those who passed. John Platt, a great American physicist from the University of Chicago, once wrote an article 'On Maximizing the Information Obtained from Science Examinations, Written and Oral' (*American Journal of Physics* 29, III (1961)) in which he argued that we should think of examinations as a mapping function. For written exams we are trying to map out the abilities of a group of students on a numerical scale. We spend a huge amount of time afterwards with our potential PhD students determining where the border is between pass and fail. But really, who cares which student is marginally pass versus who is marginally fail? What we should be trying to distinguish is the student who is really brilliant from the one who is merely good or very good. Likewise, in oral examinations we are trying, according to Platt, to map the boundaries of a student's knowledge. But those administering the oral examination rarely agree on a strategy for doing this, and instead wind up spending a lot of the allotted time exploring a very small area in great depth. Much more thought about what we are trying to achieve with examinations at all levels, from the spelling and arithmetic tests given to six-year-olds, to the tests we use to determine admissions to undergraduate education, to those given to doctoral students, is undoubtedly due. And while this has been the case for a long time, perhaps new technology for obtaining information can catalyse a rethink.

ing both the original proposition, to be sure you really understood it, and a discussion of what your thesis was about. In addition to all of this chemical knowledge, this was a time when English did not dominate the scientific literature as it does today, and there was certainly no Google Translate. So we had to pass examinations proving our ability to translate scientific articles from German and either French or Russian into English. Indeed, much of the literature that I had to understand for my thesis work was in German, so this was an important skill. And if you passed all of this, which most people who had come through the qualifying process did, then you threw a party and got on with your research project.

Teaching was the second important part of my graduate experience. The standard format was to teach first- or second-year undergraduate students in the laboratory, and in what were called recitations, where you helped the students work through problems, during your first year or two as a graduate student, and then be supported on a research grant from your thesis advisor or some sort of special fellowship for the balance of the time. I had a slightly different set-up. I did do the basic undergraduate labs the first year, but then I was asked to switch to helping Professor Robert Livingston in the teaching of undergraduate physical chemistry, essentially grading exams that he gave to the students. The first year I did this I sat in on his lectures, which were masterful. My own undergraduate physical chemistry course had been disorganised and mediocre, and while I got reasonable grades I didn't really understand it in depth. But as I listened to Livingston it was as if a fog lifted and all was clear. For some of the material, especially thermodynamics, this was the third pass through. Of course, I had to be able to work every problem that the students might face in their exams, and that helped too. Because I liked doing this, and Livingston became a mentor and friend to me, I asked to continue working with him throughout my graduate career, while having additional research grant support. And it wasn't just the physical chemistry that I learned. Bob Livingston taught me how careful I needed to be in grading, because what we did affected the lives of these students. He taught me a lot about how to construct a written examination, pointing out that the easiest exam to grade (say multiple-choice answers) was the hardest to construct (because you had to both give the right answer as well as anticipate what students would do wrong and give those answers as alternatives). Of course, having cheap labour, me, to grade led him to construct the opposite sort of exam. And he was generous with his time talking to me about what a faculty position was like, and how to be successful at it.

There was all this – exams, teaching, starting laboratory work – dominating especially the first two years of my graduate career. It was my first time living away from home, adjusting to having a roommate and doing all the domestic chores for myself as well. Still, a group of us were in it together, often sharing apartments, struggling through exams, partying when anyone was successful. But there came a point when exams and courses were over, teaching was a minor part, and there was only my thesis research to do. The lab was open to me at any time of the day or night, and I had few obligations – an appointment with my thesis advisor, an occasional seminar to attend or to present, but basically all I did was research.

Well, not all, I had girlfriends, some of whom were other graduate students in chemistry (although there were few women in the department back then), others from among the graduate students in English, occasionally a date with an undergraduate, though dating graduate students was not particularly cool. On the one hand passionate relationships are a distraction from the total immersion in science that I had as a graduate student, and on the other hand if you are twenty-two-year-old boy they are both necessary and pretty much inevitable. It is best not to have these passionate relationships within your own research group, and I almost completely avoided that.

Being a large campus there was also a variety of entertainments on offer, and I read books and had political arguments (it was after all the 1960s, and a time of turmoil of the Vietnam War and the Civil Rights movement), but from a work point of view all I had to do was my research. I think I was fortunate to realise that this was a two-year period that would be very special in my life, and I should treasure it. Every day I came to the lab, started doing experiments, and kept doing them all day, usually into the early evening. Often, my room-mates and I would have dinner together, then we would go back to the lab and work until we were too tired to do any more. Weekends were just like every other day as far as research was concerned. Sometimes a few of us went out for a meal at 2 a.m., or went to the laundromat to do our laundry in the middle of the night. If there was a movie showing, we might go to see that at 8 p.m., having first started an experiment, then go back to the lab to finish off what we had started earlier. On Saturday mornings we would go to the super-market and buy a week's worth of groceries, then maybe collect our girlfriends and take them to lunch at a bar called The Pub which served lunch for ten cents on Saturdays, maybe go to the football game if one was on, and then back to work. This was the last time in my life for at least the next forty years when I didn't have a schedule, and I could just do work that was interesting and pleasing, no meetings, no committees, no one expecting me to be there by 9 a.m. And yes, I say this is a privilege, because how many people ever get such a two- or three-year period in their entire working lives?

I mentioned that I had an 'occasional seminar to attend', and an occasional seminar to give, because part of a good graduate education is that students learn how to present their work, or the work of others, to their fellow students and to faculty, and to be able to answer questions, defend a point of view if they have taken one, deflect someone who is interrupting (usually a junior faculty member trying to show off) so that you can stick to your main point and not lose the flow of your presentation. But listening to seminars is also something to be learned, and curiously no one ever taught, or attempted to teach, us how to listen. Eventually I realised that most presentations, while they may take an hour, have just a few key points, and the art of listening is to drop away all the extraneous material and grasp what are these key points that the speaker is making. What did they set out to prove (or disprove), and how solidly did they achieve this? A lot of the rest is either experimental or theoretical detail, or, all too frequently, a description of unfruitful paths they went down before their brilliance showed them the true route to the answer.

But there is another aspect to seminars, and to conversations with visiting scientists (or, later, conversations you have with scientists you are visiting at other universities or research institutes), that will come up over and over again in what I am going to describe about my scientific life. That is using your brain as a filing cabinet. You hear things, and most of them can be discarded, but occasionally there is something that needs to be filed away because it could be useful later. Sure, you might have come across a published paper to put in an actual filing cabinet, but more often it is just a question of remembering, 'when Adam visited here, he told us about that really interesting reaction ...', and knowing when to pull that out and use it in your own research.

Far Infrared Absorption of Polar Liquids: An Unexpected Observation Needs to Be Explained, and This Opens Up the Possibility of Understanding More about Liquids

ABOUT A YEAR BEFORE I FINISHED MY PhD, I started to think about what I would do next. I had a tentative job offer from RCA Labs in Princeton, but I didn't give it much consideration – I wanted an academic job at some point, and that meant doing at least a year of postdoctoral research. If I had been better advised, by either Walter Edgell or other faculty at Purdue, they would have told me to go to Berkeley or Princeton, MIT, Harvard, someplace with the maximum amount of prestige, because it was from those places that the best academic jobs were being filled. Professor L. B. (Buck) Rogers at Purdue, who led the analytical chemistry division, and with whom I had done a small side project during my PhD work, told me to look for a job as an analytical chemist rather than a physical chemist – there was much more demand for analytical than physical at that time, and I could do the same sort of research. That was good advice and I just ignored it. Silly boy.

My old lab mate, Stan Abramowitz, from my undergraduate days at Brooklyn Poly, had joined the staff at the National Bureau of Standards in Washington, DC, and he and other friends held out the opportunity of a postdoctoral position at the National Institutes of Health with a very good scientist named Ted Becker. This could be a multi-year position, and it was attractive in that it was well paid, and mine for the taking, plus I knew Ted and his colleagues and liked them very much. The only problem was that it was not really a good route to an academic job.

In my thinking about all this, I somehow began to look at the possibility of going to Europe for a year. I had never been out of the US, even as a tourist, and wanted to have that experience. From a language point of view, the easiest would be to go to England, and in my field of interest there was Professor Norman Sheppard, who was well regarded and at Cambridge University. But a problem arose immediately, namely that he was moving to a Chair at a new university in East Anglia at exactly the time I would be coming. Now even I realised that to join that group would be a bad idea – not a known university, and the whole lab had to be set up from scratch, so the chances of having a productive year of research seemed slim.

When I talked this over with my undergraduate advisor, Bob Bauman, as we strolled around the campus of Ohio State University during a meeting of spectroscopists, he told me that he had just completed a trip around several European spectroscopy labs (Bob's idea of a summer vacation), and his highest opinion was of the lab of Professor Hans Gunthard at the ETH, the Swiss Federal Institute of Technology, in Zurich. Very well equipped, an outstanding place for science, and Gunthard himself a great scientist to work with. This was very attractive to me, and a little background reading (this required going to the library, as the internet had unfortunately not yet been invented!) revealed just what a great scientific centre Zurich was, and the ETH in particular. Every professor of organic chemistry there had won the Nobel Prize.

If I was to go to Zurich and work in Gunthard's lab, I wanted to have my own money, so that I would have the freedom to work on a problem of my choosing, and that meant securing a postdoctoral fellowship from the National Science Foundation or from NATO, the two sources at that time who would consider an international fellowship for an American. At the same Ohio State meeting, I was pointed to someone who knew just how to secure such a fellowship (he had probably sat on the panel that made the selections one year), and he generously told me what the key things were that I needed to say in my application to make it compelling. So we go to scientific meetings to learn new science, but also to make friends, and get help with our careers. This was the first of many such instances of my receiving help from a conversation at a scientific conference.

In the Purdue chemistry department, Professor Nathan Kornblum had been my teacher in several courses in organic chemistry, was a member of my thesis committee and always took an interest in my career. I knew that he had spent a sabbatical year at the ETH and asked his opinion of it as a place for a postdoctoral year. He was super enthusiastic, felt it was one of the best places for science in the world and as good a place as I could possibly pick. If I had any doubts, this convinced me.

I wrote to Gunthard, introducing myself and my background, and told him what problem I wanted to work on if I could join his lab. He immediately responded with an invitation. With that in hand, I applied to the NSF for the fellowship using all the key phrases I had been taught. As it happened, though I didn't know it, of course, one of the Purdue physical chemistry faculty, C. R. Mueller, who had been my professor in thermodynamics and statistical mechanics, was on the panel that year. One day in early December he spotted me in the corridor and said that while he had abstained from voting on my application, as was proper, I had a very high ranking among the applicants and was sure to get it. And in due course that was what happened.

So that is how (newly married, but that is another story) I arrived in Zurich in the first week of July 1966. I had two years of college German, and some French, and I just figured that somehow I would get by on the language front. That was true, but it was very difficult at first. The guys in the lab spoke English very well, but when they spoke in German it was Swiss German dialect, of which I could understand not a single word. Because Gunthard believed that it was crucial for all his PhD students to be fluent in English, all of our group meetings with him were conducted in English, and would have been even if I had not been there. Still, going to departmental seminars was difficult at first, unless the speaker was from England or America. And, of course, I needed German for everyday life: shopping, dealing with the landlady from whom we rented our apartment, discussions with neighbours and every now and then with the police when we went through the formalities of registering our presence. But it is a funny thing with languages, at least when your brain is still a bit plastic. I started to attend evening classes once a week, which was also a place to make friends, and then I just listened. Gradually I started to understand what was being said, and I worked up the courage to speak as well. My wife, Elly, started to play basketball with the Zurich women's team and got to speak with them and travel around with them, and she also went to classes. It took about six months before I could follow conversations, but when I left Gunthard said that he thought I was understanding about 90 per cent of what was said even in Swiss German.

As to the science, the problem I had picked was this: one of the regions of the spectrum that was being newly explored, indeed that I had done some work on myself both as an undergraduate and graduate student, was what was known as the far infrared, the lowest-energy part of the infrared region. It had been a challenge because of instrumentation difficulties, and these would not be really overcome until some years later, but it was possible now to make good-quality measurements on the first commercially available instruments. Bob Jakobsen and Jim Brasch at Battelle Laboratories had published a paper which caught my attention. They had taken a whole series of organic liquids and just measured their spectra in this region, and found that the polar liquids, that is, those containing certain kinds of bonds between different atoms, like carbon–oxygen, sulfur–oxygen or carbon–nitrogen (called polar because one atom attracted the electrons, negatively charged, from the other, so that the bond looked like it had negative and positive 'poles') absorbed radiation over a broad region of the far infrared, whereas other liquids, pure hydrocarbons for example, did not. They called this a non-specific absorption, since it seemed to be general to a broad class of compounds. But what could be causing it was unexplained. A few other papers along the same lines had appeared, including one from the laboratory of Professor Gunthard, and each of these proposed a different possible mechanism, but without giving any proof for their hypothesis. I thought this deserved further investigation and explication, and that if I could design a proper set of experiments I could figure it out. Fortunately, Gunthard's lab had just acquired the newest instrument, the Perkin-Elmer 301, to make such measurements.

This problem was another aspect to the completion of the 'painting' of chemistry. I have said that the new techniques of spectroscopy allowed scientists to solve longstanding problems of structure and reactions. But in the course of using these techniques a number of observations appeared that were unexpected, in effect creating some new problems. By explaining these things in terms of what we already knew about chemistry, the power of the techniques increased, while our understanding of such things as the nature of liquids became more profound. In effect, chemistry set out to solve problems of structure of molecules and what happened in chemical reactions and expanded to an understanding of the properties of solids, liquids, liquid crystals, polymers, gels and gases.

Now the time is 1966. Universities had computers, of varying degrees of capability, and I actually knew how to write programs to make use of these computers. This was entirely thanks to John Lundberg, whom I have already mentioned as a visiting professor at Brooklyn Poly during my senior year. Poly had an early IBM computer, the IBM 650, which you had to program in a very simple machine language. Most of the chemistry faculty were not yet using this computer in 1961, but Lundberg insisted that it was the future, and every one of us who were at the start of our scientific careers should learn to program, which we did. Even five years later, in 1966, when I was in Zurich, the computers were big hulking things made by IBM or a company called CDC, housed in the computer centre someplace on campus. To input a program or data into such a computer, you had to punch cards and then carry those cards to the computer centre, where, when it was your turn, the cards would be read into the machine and your program run. Output appeared on large sheets of paper from the printer. And if you made an error, which of course I did quite a few times, it was back to the beginning again. If you tripped on the street on the way from your office or lab to the computer centre, and the cards spilled all over the wet pavement, well ... I did that too.

At the other end, our laboratory equipment for recording data was all analogue, and there was no computerisation of the instruments. That severely limited the processing that could be done on the data, hence had a big influence on what experiments one could design to solve a problem. In the case of my experimental design to understand the non-specific absorption in the far infrared spectra of polar liquids, I needed to do some precise measurement of the intensity of the absorption, measured by the area under the band that I recorded. Now at that time the only ways to measure the area were to use some sort of a tracing device that you rolled over the measured band, and this provided an approximation of the area, limited by the difficulty of doing this tracing reproducibly, or, and this was fun, cutting out the measured trace and weighing it, assuming the paper had uniform thickness and density. This was also not very good.

When I became familiar with the facilities in the Physical Chemistry Institute, which Gunthard led, I found that we had at our disposal not just great instrumentation for measurement, but the very best machinists, electronics technicians and even computer programmers to help us. And in abundance. One small problem: while the graduate students and researchers all spoke English, and Professor Gunthard believed that English was an important skill for his scientists to perfect if they were to be successful on the global stage, the machinists and technicians only spoke German. Well, this was as good a motivation as any to improve my German.

My hypothesis about the far infrared absorption was that it was associated with the interaction of polar groups in the molecules, vibrating against one another between molecules. For example, if you had the molecule CH_3CN, known as acetonitrile, two molecules might line up as shown below, and there could be a vibration of the molecules against one another, something like indicated by the arrows. But because these associations of the molecules are much more poorly defined than actual chemical bonds, the vibrations do not occur at a sharp, well-defined energy, but more broadly. And this was consistent with the observation.

To test this, I proposed two things. First, that if you had a molecule of general formula RCN, where R could be CH_3, as shown below, or CH_3CH_2 (or more generally $CH_3(CH_2)_n$, where n was o, 1, 2 ... etc.), as the carbon chain became longer the associations between the CN groups would become weaker, and the absorption in the far infrared would decrease. One way of looking at this is that when n = o or 1 the molecule is dominated by the CN part, but as R increases the molecule becomes more hydrocarbon-like. And if one picked something like the C_2 or C_3 hydrocarbon chain, and dissolved the molecule in a hydrocarbon solvent, as it was diluted

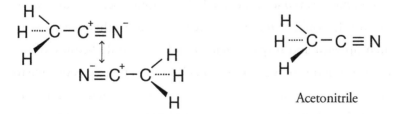

Acetonitrile

more and more you would break down the possibility of association, and again the absorption would weaken.

But to do the latter experiment, the dilution, you need to keep the same number of molecules of RCN in the path of the infrared beam, so that as you diluted the RCN, the possibility of absorbing the radiation stayed constant. So if, for the pure liquid, the cell in which the liquid was contained had a thickness of 0.05 mm, when it was in a 10 per cent solution the thickness needed to be 0.5 mm, etc.

And, of course, these liquids all have to be very dry, because water absorbs strongly in the far infrared region of the spectrum and would distort all the results. But I knew how to do that from my graduate work.

Now the experimental design required that I be able to measure the total amount of light absorbed very accurately. To do that, I decided to modify the instrumentation so that instead of providing the output to a pen on paper, it made the measurement at each frequency and punched the result on a card. This required an electro-mechanical device called a stepping motor to change the frequency in digital steps, rather than analogue continuum, and an electronic device to capture the absorption information. Both of these then had to be translated into a format suitable for a card punch and sent to the punch, which was a big noisy device to have in a laboratory.

It took nearly three months of effort to design and build this system, but it was one of the first fully computerised instruments of this sort. It would be more than five years before others were able to catch up with and surpass what we did there, only because of the support in terms of machining and electronics that was available to me. For the next eighteen years in academic research, I never had this quality of support; indeed, it is available in very few universities in the world, and this limits the type of research that can be done in university laboratories.

The results of this work were published in the journal *Helvetica Chimica Acta*, a Swiss chemistry journal, quite respectable. Professor Gunthard generously said that I should have it as my sole work, though convention would have had him as a co-author, but he knew that having a publication with only one name on it was good for someone starting out his career. When I reread this publication now, after many years, I see how poorly written it was. There were results whose conclusions I did not clearly state, the abstract didn't tell what the point of the article was – just a generally amateurish job. But it was the beginning, the first thing I had done completely myself, and there was one graph in there that made the point – after that there was not much need for others to re-investigate this so-called non-specific absorption.

Vibrational Spectra of Liquid Crystals: Probing of 'Phase Transitions' Becomes Possible, and This Can Be Applied to Biological Membranes

WHEN YOU ARE STARTING OUT AS AN assistant professor, it is very important to pick an area of research where you can attract some funding, get results, be noticed and, of course, make a substantive contribution to science. Unlike the three problems I had already worked on, which were best because they were narrow, well-defined issues where I could get a definitive result in a short time, I had to find an area of research that had the scope for a number of different problems that my prospective students could work on and that could lead to a body of publications around which I could build my reputation as a scientist. Now I had a background and interest in spectroscopic techniques, and my interests were in liquids and solids, and how they were structured. In effect, the work I had done during my year in Switzerland gave me a little grounding and a tiny reputation in this. While I was doing my experimental work in Zurich, I also had lots of time to study theory of liquids, and to look at where there were interesting problems.

At that time, mid-1960s, there was a revival of interest in a class of materials known as liquid crystals. Liquid crystals, a phase of matter having the flow properties of a liquid but some of the order of a crystal, were the subject of a lot of studies in the 1930s, then seemed to fade from scientific interest. In 1962, Professor George Gray of the University of Hull published a book entitled *Molecular Structure and Properties of Liquid Crystals*, the first book in English summarising all of the research results and known techniques for studying liquid crystals, and this revived interest in the field. As a result of George Gray's book, a number of academic and industrial laboratories started to study liquid crystals in the 1960s, notably the RCA Sarnoff Research Labs in New Jersey, where some of the initial work on using liquid crystals as displays for electronic signs was done, the laboratory at Collège de France in Paris and at Orsay, with Professor Pierre-Gilles de Gennes, one of the twentieth century's greatest physicists, and also near Paris the laboratory of Vittorio Luzzatti and Annette Tardieu, who saw the possible role of liquid crystalline phases in biology.

So in 1966–7, when I wrote proposals for funding in this area, I was at or near the front of a wave of interest. My work had two thrusts, both of which used infrared and Raman spectroscopy to probe the liquid crystalline phases. The first was with the most well-established organic liquid crystalline materials, one of which was known by a shorthand as PAA. As in my work on polar liquids in Zurich, this material had methyl (that is, CH_3) groups at either

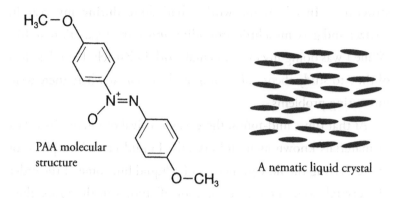

PAA molecular
structure

A nematic liquid crystal

end of a complex structure, but it also existed in homologues that had longer hydrocarbon chains just like the RCN molecules I had studied, and these were also liquid crystalline materials. For these materials, there was a well-defined crystalline solid; then, if they are heated, at a definite temperature (or so it appeared) the crystal transformed into what was called a nematic liquid crystalline phase, shown schematically above for the rod-like molecules of PAA, which to the naked eye appeared as a cloudy yellow liquid; then at another definite temperature it became a clear liquid. I wanted to build on my work that I had done in Zurich to see how the vibrations of the lattice (that is, the vibrations of the entire molecules against one another in the solid or liquid crystal) changed as we went through these transitions.

The second thrust was to look at the biological molecules that formed the structure of cell membranes, for example in red blood cells. These are a class of molecules known as phospholipids, and when they are mixed with water, under certain conditions, they form a phase which also appears to have liquid crystalline properties. This was of possible interest in several different directions for medicine. If the permeability of the membrane to either drugs or other poisons for the cell changed depending on the liquid crystalline properties, then that could have implications for both the cause and treatment of cancer. And it was already known that cholesterol bound to the membranes, but just how it did, and whether that binding greatly modified the fluidity of the liquid crystalline phase, or even stopped it from forming altogether, was unclear. Obviously this was crucial to aspects of heart disease.

My first problem was to get an academic job, at a university where I could do research. At that time the American Chemical Society published a regular bulletin called Academic Openings, and I applied for whatever openings there were that appeared to suit my background and interests. I got some nibbles, from George Washington University in Washington, DC, State University of New York at Stony Brook, Hunter College of the City University of New York, and Villanova University in Philadelphia. I was also asked to come and interview for a position which would be purely research, no academic connection, at a new laboratory that NASA was opening in Cambridge, MA, adjacent to the MIT campus. So in due course I took two weeks off from my work in Zurich and my wife and I returned to the US for these interviews. At each place I gave seminars on my work, and for the most part they went very well. I can't recall what happened at George Washington University, but it was not a very inspiring place at the time. At SUNY Stony Brook, a new campus in the early stages of staffing up, I was inspired, but there was a professor there, Harold Friedman, who was very hostile to what I was doing, and seemed to feel I was not up to his standards. I am sure he vetoed any consideration of me for a faculty position. Villanova was very cordial, and offered me a position as assistant professor, but it was also a bit uninspiring in terms of where the department was going. But at Hunter College, an old institution, there was a change under way from purely undergraduate education to a fully fledged undergraduate and doctoral programme as part of the City University of New York, with many new faculty being hired. Several senior people had come, among them Dick Wiley, an established organic chemist who was leading the development of the doctoral programme, and Jeff Wijnen, a physical chemist with a fine reputation. Wiley was particularly interested in the possibility of my establishing a laboratory to do Raman spectroscopy. There were also some recent junior faculty who had been hired, among them Maria Tomasz, who was to become a great biochemist, and Bill Grossman, an analytical chemist with interests somewhat related to my own, who had come from Cornell and a postdoctoral fellowship in England. The place had no reputation for research, but it was in New York City, which worked well for my wife, who wanted to enter the NYU graduate programme in American Studies, and Hunter seemed determined to build something. So when they offered me a position as assistant professor I was inclined to accept.

The NASA lab also offered me a job, and this was a tougher decision than turning down Villanova. In my field, to do work you need state-of-the-art instrumentation, and NASA had bought it, all of it. They already had the instrument that everyone in my field wanted to have, with every possible accessory. But there was something wrong. As I spent my two days there, giving my seminar, talking to the management of the laboratory and other scientists who were already there, I developed the feeling that they didn't have the slightest idea what to do with all the equipment they had bought, or why this was relevant to the mission of NASA. It turned out the lab had been created and placed in Cambridge because of a political intervention by Senator Edward (Ted) Kennedy, who had been elected to the US Senate a few years earlier to replace his brother. NASA didn't want the lab, and the lab really didn't want NASA. There was just something vaguely unsatisfying about the whole thing. So despite the fact that I would have had lots of equipment, and a salary about 50 per cent higher than what I was offered by Hunter, I turned them down.

Now I had an academic job confirmed, and I could start my independent scientific life.

While I was still in Zurich I wrote a proposal to something called the Petroleum Research Fund of the American Chemical Society, which gave small ($5,000) so-called starter grants to new faculty. These were hotly competed for, as virtually every new chemistry assistant professor who was expected to do research applied. I had to have some referees, and Professor Gunthard, my postdoctoral advisor, was an obvious one. He asked me what he should say, and I told him that this was a competition with very few winners, so it was necessary to be extreme in his praise. He got the message, and wrote something like 'the research proposed herein is among the most important in physical chemistry, and if successful would likely be a candidate for the Nobel Prize'. A bit of an exaggeration but it seemed to do the trick, because I got the grant.

This gave me a little start, and because I was a new faculty member at Hunter College of the City University of New York, and none of the new faculty either at Hunter or at other campuses of the City University had received one of these grants, they were excited. Dick Wiley took me to see Ruth Weintraub, the graduate dean at Hunter College, and Mina Rees, president of the CUNY Graduate Center, and each gave me additional money, in effect tripling the size of the grant. This allowed me to buy a few key pieces of equipment to make infrared spectroscopy measurements at carefully controlled, variable temperatures. Now I could start, and in just a few weeks I had my first student, Dolores Grunbaum.

But I needed more. What I wanted was an instrument to do Raman spectroscopy. In 1964 lasers suitable for Raman spectroscopy became commercially available, and two companies, Jarrell Ash and Spex Instruments, built Raman spectrometers incorporating these. I wanted to apply for money from the American Cancer Society, because they had a mission to support a small amount of basic science along with the more targeted medical research, they were local to New York City and their review cycle was short. I knew what I wanted to write about, and what I could ask for, but the maximum-size grant they gave was not enough to both support the students I needed to make measurements and also buy the new instrument.

Now I got two good pieces of advice. The first came from my colleague Maria Tomasz. She was a few years beyond me as a faculty member and had already received a grant from the National Institutes of Health. I asked for her help, and she told me that the problem with many grant proposals is that they talked about an interesting area, but didn't really say what the investigator was going to do. Her advice was to say this very explicitly, something like: 'To attack the problem we have laid out, we are going to do the following set of experiments. If the results are A, it will mean this, and will lead us to do the next set of experiments. If they are B ...' In other words, and this might seem obvious but it isn't to most scientists, just send in a detailed research plan and you have a very good chance of success.

The second important piece of advice came from a man named Ernest Bergauer, who was in charge of the process for sending out research proposals from Hunter College. I told him about my problem with not being able to ask for enough money from the American Cancer Society to do what I really needed to do, and he asked how much I was short. Well, about $10,000, I thought. 'OK,' he said, 'we are going to get three different deans to each commit to give you $4,000 if you get the grant, so you will have an additional $12,000. Now they will commit to this, when I ask them, because they will be sure that you will never receive the grant so they can promise anything. I will get these commitments from them in writing, and lock them up in a safe, because I can tell you that if we don't have that they will certainly try to deny that they made the promise. Moreover, you will write these matching commitments into the proposal you send, and this will be signed by the college provost, who will not pay the slightest attention to what he is signing. Good luck!'

Well, I did get that grant, and in fact the American Cancer Society supported my research for most of the eight years I was at Hunter College. Later, using the same systematic approach to describing what I planned to do, I got a large grant from the US Army Research Office. This was controversial, because the Vietnam War was raging and New York City, including Hunter College, was the centre of various protests against the war. One of the faculty members in the chemistry department, Charles Hecht, was particularly virulent on the subject and active politically. When he found out that I got a grant from the army he never spoke another civil word to me. I knew nothing about army research, but once again they had a mission to support basic research and somehow I succeeded. And I got two good grants for additional instrumentation, one from the National Science Foundation. Because of all this, I was able to attract students and do the work I wanted to do, even though I was not at a first-rate graduate school. Because we were in

New York City, I had some very fine students to work with me, better than we deserved to have. The results of our work, about which more below, were published in some of the best journals.

During this time my parents were still living in New York City. In the early 1970s my father retired and they moved to Israel. My parents had no scientific education at all, and had no idea about how science was done, but they believed in it, and they certainly believed that it was a good thing that I had done a PhD and had an academic job, for lots of reasons, but among them because it meant that I would not be drafted and sent to Vietnam, a real risk at the time. When I received my first big grant from the American Cancer Society, they were sure that it was only a matter of time before I would get a Nobel Prize, and this was communicated to my many aunts and uncles.

I think the same naivety or ignorance of what we did in science, how we did it, what were the rewards, was common to most of my colleagues. None of us seemed to be second-generation professors. I recall my colleague Ed Abbott saying how he had proudly told his father that he had just published his first paper in a scientific journal, and his father asking, 'How much did you get paid for it?' And of course we had to pay (or our research grants paid) a page charge for the articles, to support the journal. So the answer was something like -$500!

So now to return to the problem of learning about liquid crystals from their infrared and Raman spectra. When we examined the infrared absorption spectrum of PAA as a thin crystalline film, we found that it had a number of very sharp absorption bands that were not present in other phases. These were separated by a small energy difference from the main bands that appeared in the liquid or even liquid crystalline phases. I was very pleased to see these, because I was sure that the only explanation was that these arose from a strong coupling of the vibrations of the molecule with the vibrations of the lattice, that is, of the whole molecules vibrating against one another. These 'side bands' are not usually seen, but for a material which is going to form a liquid crystal, the lattice forces binding the molecules together must be very strong. A quick experiment showed that they all disappeared in the liquid state, and in solution, as well as in the liquid crystalline state itself. When Dolores heated the sample and measured the intensity of these bands as the crystal–liquid crystal phase transition was approached, she found, to our surprise, that they did not disappear abruptly; rather, there was a decrease in their intensity over a broad temperature range before the phase transition, after which they were gone. And when we looked at the homologues of PAA with longer hydrocarbon chains on the ends, the shape of the pre-transition effects was different, the transition being more abrupt with the shorter chains.

Now I knew from my graduate education and from reading Thomas Kuhn's *The Structure of Scientific Revolutions* that it is very important when doing scientific research to not expect a surprising result. Every good experiment should have a hypothesis, usually based on the prevailing understanding in the field. (This is, incidentally, the complete opposite of invention, where you are only interested in something that is unexpected or not obvious.) In this case the wisdom said that the crystal–liquid crystal transition was what was called first-order, meaning that it occurred abruptly, with no pre-transition effects. A few studies had indicated otherwise, but they were inconclusive. We would have been happy finding a first-order transition, and showing that our extra bands were indeed monitoring the transition. But the data said otherwise.

To prove that what we were seeing was associated with a crystal–liquid crystalline phase transition, and not some anomaly of our measurement technique, we looked at similar molecules that did not form liquid crystals when they melted. In some of these, we were also able to find these sharp bands, but we demonstrated that they disappeared abruptly rather than gradually. This is another part of the scientific method, using a negative result to support your hypothesis.

To put some further analysis behind this, having looked at six different homologues of PAA, we went back into some classical physical chemistry that I had learned as an undergraduate and combined it with some ideas from work on liquid crystals by the great chemist W. L. Bragg in the 1930s. Based on Bragg's work, we proposed a model for what was happening, and using the undergraduate physical chemistry ideas we were able to calculate energies for breaking a molecule free from the lattice. When we computed these energies going up the series from 1 carbon to 6 carbons, the energies decreased, but not smoothly, rather they alternated, so (the units are unimportant here) C_1 was 42, C_2 was 128, C_4 was 10.6, C_5 was 19.6 and C_6 was 16.1. George Gray, in his book, had already noted an alternation in other properties and had even given an explanation for it. Now we provided some further proof that his explanation was likely to be correct.

When we did this work, we did not yet have our Raman spectrometer available. After that was installed and working, we went to look directly at the lattice vibrations of PAA. We were lucky in that PAA turned out to be a very good 'performer' in the low-energy Raman spectrum – at that early stage of such instruments not everything we tried yielded high-quality data. Once again we developed equipment for varying the temperature in small steps, and this worked better with Raman than it did with infrared because we could use much smaller sample sizes, as the sample was in a focused laser beam.

As a result, temperature uniformity was much better, and we could see more subtle changes.

Sure enough, as we approached the crystal–liquid crystal transition, a few of the lattice vibrations moved across the frequency range towards zero energy, while others did not. Then once we were in the nematic liquid crystal there was a broad scattering associated with this ordered fluid, much like what I had studied back in Zurich for polar liquids. But this movement in frequency across the spectrum was an important observation. From my study of the basics of solid state physics as a graduate student, I knew that there was a class of inorganic materials known as ferroelectrics, where it had been discovered that as they approached a phase transition there were vibrational modes that did this, moved to zero frequency, and then reappeared on the other side of the transition. These were called soft modes. And now we had observed them for a liquid crystalline material. This was consistent with our earlier view that the crystal–nematic transition was not first-order. Unknown to me, Professor de Gennes was exploring very similar ideas in Paris, but with much greater depth of understanding and originality of the physics involved.

As usual, to do first-rate science rather than second-rate, we needed to put some theory behind this, or at least some more detailed calculations. We grew single crystals of PAA, about a millimetre on a side, and highly ordered. We aligned these in an X-ray beam, so that we knew which axis was which – how the molecules were facing in the crystal, if you will. Then we transferred this crystal to the Raman instrument, changed characteristics of the laser beam and were able to both isolate each of the lattice vibrations and prove which one it was, i.e. to which movement of molecules it corresponded. While this was something that others had been doing at around that time, no one had done this for such a complex molecular crystal before.

Having made all the spectroscopic measurements, I decided to try to calculate the entire lattice vibrational spectrum. There were parameters in the literature about how parts of molecules are attracted or repelled by one another, and we realised that these parameters could be used the same way that spectroscopists were using 'force constants', a measure of the strength of chemical bonds as they vibrated, to calculate the frequency of the bands in infrared and Raman spectra. We regularly did such calculations in Edgell's group when I was a graduate student, and I also wrote a computer program to do them when I was in Zurich. Dolores constructed a very complex computer program to use these parameters to calculate the vibrational frequencies for the lattice vibrations, trying to understand more quantitatively the experimental data we already had. When I spoke about this work, later on, I would compare it to the work of one of my colleagues at Hunter in the Art Department, the great minimalist Tony Smith. One of his greatest achievements is a large sculpture on the campus of Bennington College, and one of the key things about this was that he didn't actually make it himself, he called a foundry and ordered it. Well, that is what we did in the calculation, we just ordered up the parameters for the calculation from the literature, and put them together. So now we had the phenomenon, observed in both the infrared and Raman spectra, and a series of measurements and calculations to back up our explanation of what we had observed.

The results appeared in a series of papers in the *Journal of Chemical Physics*, a prestigious place to publish. I proudly titled the first one 'Vibrational Spectra of Liquid Crystals. I. Changes in the Infrared Spectrum at the Crystal–Nematic Transition'. It appeared in 1969 and was the first of fifteen papers in this series. If you are going to build a reputation in science, you need to produce a body of original work in an area, not just some random shots here and there. By making it clear that our work was a series, something that was more fashionable then than it is today, I could announce to the scientific community that this was my intent.

We published our calculations, and we did far infrared spectroscopy of the molecules, completing the picture. Dr David Beveridge had joined us at Hunter by then, and he had developed methods for doing structural calculations of complex molecules. With help from Dave, we did just that to be sure we knew the subtlest aspects of the equilibrium structure of the molecules we were studying. In all of this, I was influenced by a talk I had heard Frank Bovey of Bell Labs give, when he spoke about why they did so much detailed study on certain polymers. The Bell system was the biggest user of these polymers, as insulation on cables, and it was simple – they wanted to know everything there was to know about them. So if I was building a case around the nature of certain liquid crystals, I too wanted to know all there was to know.

Now at Hunter, at all good universities, there is a regular programme of visiting speakers, and the routine is that you spend half an hour with them, tell a little about your research, exchange ideas, etc. And have a nice lunch as well as hear their talk. We did a lot of this at Hunter, because no one knew who we were, but everyone fancied a trip to New York City, and after their visit to us they got the idea that we were a good group of young researchers. Hunter is located on the East Side of Manhattan, 68th St and Park Avenue, so as you can imagine there are lots of good restaurants around. Joe Dannenberg, one of my colleagues, discovered a place at 3rd Avenue and 72nd St called La Bourgogne, run by two young French guys. It was on the ground floor of a luxury apartment building, and many of the elderly ladies who lived there would come down to lunch together. When we started going to La B, as we called it, you could get a very good three-course fixed-price lunch for under $5! Add to that some Muscadet or Beaujolais and it was possible to entertain visiting speakers very well indeed. Some of them imbibed a bit too much and were barely able to deliver their seminars by 4 p.m.

One day, Professor Rudy Marcus from Caltech was visiting to give a seminar. I have mentioned him before as having been a faculty member at Brooklyn Poly when I was an undergraduate, but by now he was at Caltech. During my discussions with him, my half hour, Rudy Marcus mentioned some ideas to me about how one could study motions in liquid and perhaps in liquid crystalline phases, and I took up these ideas. Here the measurement side was fairly easy, but what was difficult was the mathematics of extracting the information from the measurements. When I was a graduate student at Purdue, we were required to have a major and two minor subjects, even as graduate students. My major was physical chemistry, and that is where I took most of my courses, one minor was organic chemistry, probably not that relevant to my later career, but my other minor was mathematics. Now as an undergraduate I hadn't been all that outstanding as a maths student: five semesters were required and my grades were a bit up and down. But I pushed myself to do more in graduate school, and took four graduate mathematics courses, two terms of what they called Advanced Engineering Math, and two of Advanced Calculus. Yes, it was hard, and yes, I used more than half of what I learned in my later career. I once read a short autobiographical book by the great astronomer Fred Hoyle in which he said that to do great work in astronomy you can't be afraid of the mathematics; sure, it is easier to be qualitative and avoid learning the maths, but you will never tackle the really hard problems. The same was true of physical chemistry.

We took this on and produced a paper that was able to show quantitatively for the first time just how the motion of molecules changed as one went from liquid crystalline phases to liquid. By the time we published this in the *Journal of Chemical Physics* in 1978 it was paper XI in the series. At one point we thought that our work had led us to discover something new in mathematics in an area known as Fourier Transforms, the mathematics that relates things changing in frequency with changes in time. But how could we know if it was new? We just didn't have enough background. Fortunately, by that time I was working at Polytechnic Institute, and in the Electrical Engineering Department was Professor Athanasios Papoulis, perhaps the world's authority on the Fourier Transform. So I said to my student, we are going to show this to Papoulis and see what he says. Off we trekked to his office with our papers. Papoulis sat there and looked at it and looked at it, his face changing from curiosity to interest to a smile. Finally, he looked up and said, 'Well you certainly have used an amusing notation, but I think you will find that this is exactly what is on page 278 of my book on the Fourier Transform.' Well, that saved us the embarrassment of publishing it as something new and gave us the warm feeling of having discovered for ourselves something in mathematics that was already known.

All of this work on simple organic liquid crystals was excellent science, published in the best journals, but it was of no interest to the American Cancer Society. In parallel with these studies, I had a group of students working on the phospholipid-water gels that formed the structural element of cell membranes.

These molecules have a polar head and a hydrocarbon tail. If you mix them with water, they form a gel, with the water binding to the head, and the tails trying to stay as far away from the water as possible. A way that they can accomplish this is by having the tails face inward in a two-layer, or, as it is known, a bilayer configuration, with the heads out on either side. Then if these bilayers were to form a sphere, you could have an aqueous solution on the inside and the outside, but the ability of anything to move into or out of the centre of the sphere would depend on its being able to pass through the hydrocarbon bilayer. And such spheres were not a bad model for certain biological cells.

This was becoming known science at the time I started working, and many people, from medicine to physics and everything in between, were paying attention to these lipid bilayers.

With our spectroscopic techniques, I was able to see how the hydrocarbon chains of the lipid bilayers went from fairly solid, to liquid crystalline, to liquid as we raised the temperature. Then we added small molecules to the lipid bilayers, and induced the fluidity without changing the temperature. So we had a sensitive probe of the interior of the membrane.

This work was published in the *Journal of the American Chemical Society*, and some of it was further pursued when the first postdoctoral fellow joined my lab. But we did not have the medical or

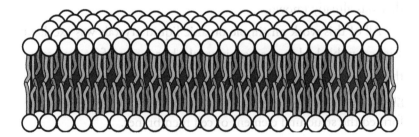

biomedical colleagues that we needed. A bit of outreach with Sloan Kettering Medical Center nearby enabled us to actually measure the spectra of red blood cells, but the work, which was published, turned out to be incorrect. A few years later, a colleague at another university repeated our measurements and saw that what we had thought we were measuring with the cells was actually due to something else that had been added to the preparation by the guys at Sloan Kettering. They just neglected to tell us, and we didn't know the right questions to ask. It happens, you make mistakes, they get corrected, and it is not the end of your career. Generally, it is better to try out some of these conclusions in a talk at a conference, because with 30–100 experts in the audience I have found there is a good chance that someone will ask the right question, and save you the embarrassment of being wrong in print. But in this case it seemed too straightforward. The referees who reviewed the paper for the journal *Biochimica and Biophysica Acta* didn't see the problem, and it was published.

We had a good time with our model membranes, and they were a fruitful source of data and publications. But it was left to others to take up our initial work and really apply it to biological systems.[*]

---

[*] The work on liquid crystals in my lab at Hunter College was done by graduate students Dolores Grunbaum, Namby Krishnamachari, Ken Brezinsky, Robert Hauser and Rose K. Rose, undergraduates Frank Prochaska, Art Noguerola and Wai Bong Lok and postdoctoral fellows Nehama Yellin, John Lephardt and K. Krishnan.

11
Na
Sodium

Some Thoughts on Understanding Condensed Matter

P HYSICAL CHEMISTS AND CHEMICAL PHYSICISTS (AN AMUSING disciplinary distinction, I am sure you will agree) talk about phases other than the gas phase as condensed matter, so this includes liquids, colloids (in which an insoluble substance is microscopically dispersed in a liquid, and does not settle for a very long time), gels (essentially colloids that look semi-solid in appearance), liquid crystals and solids. Metals are a special case of solids, and semiconductors another special case.[*] Physicists are also interested in more exotic phases, having special electrical or magnetic properties, some at very low temperatures or very high pressures. This

---

[*] Most solids are insulators, that is, they do not conduct electricity, or in electrical terms we say they have a very high resistance. Metals conduct electricity, and in the way we now think of them, electrons are free to move through the metal, so they have a low resistance. As temperature increases, the conductivity of metals decreases. Semiconductors have resistance to electrical conductivity, but it is much lower than that of insulators, and as temperature increases their conductivity also increases. And then there are superconductors, materials that, usually but not always at very low temperatures, show zero electrical resistance.

interest is due, in part, to the device applications of such materials, but also to the fact that properties such as superconductivity cannot be explained without invoking quantum mechanical effects beyond those needed for understanding metals and semiconductors.

So while through the second half of the twentieth century physicists were advancing their understanding of the more exotic phases of condensed matter, chemists concentrated on a better understanding of liquids, gels, colloids and, from the late 1960s onwards, liquid crystals. It is in these phases that most non-gas-phase chemistry occurs. Chemistry is almost always about rearranging electrons from one molecule to another, or transfer of electrons between ions, and so the electrical properties of the condensed phases in which the reaction occurs are crucial to the rates of reactions, and sometimes to the products that are formed. In parallel with all the work on reaction mechanisms that I described earlier, physical chemists used spectroscopic and other measurements, plus a lot of calculations, to understand the electrical properties of liquids and how these influenced reactions. Interestingly, the most common liquid, water, is one of the most complex as a reaction medium. Moreover, chemists were the ones who realised that liquid crystals, because they had interesting electrical and optical properties, could be used to make display devices, now part of many televisions.

There is a vast literature on these condensed phases. Many of the basic laws were laid down in the eighteenth and nineteenth centuries, in part by chemists trying to provide a scientific basis for such industries as brewing beer and making wine. While some of the fundamentals behind these laws required quantum mechanical explanations, a perhaps more universal theoretical formulation was that of statistical treatment of the phases, using techniques that are under the broad heading of statistical mechanics. This developed at about the same time as the beginnings of quantum mechanics, at the start of the twentieth century, led by Boltzmann in Vienna, Maxwell in England and earlier by Gibbs in the US. Gibbs laid the foundations for understanding all of thermodynamics in terms of statistical behaviour of the very large numbers of atoms/molecules (though he did not really know of molecules) in a system. So the theoretical basis for much of our understanding of condensed phases of interest to chemistry was there at the beginning of the twentieth century, but it took some time to overcome opposition to the ideas, then to develop the theory towards a methodology for calculating properties of condensed (or, for that matter, gaseous) phases using the statistical mechanics approach, in parallel with that, for systems of interest, bring together quantum mechanical understanding of the energy levels with the statistical view of how energies would be distributed among molecules at a given temperature, and finally for the calculations to be doable as a result of the availability of computing power. This was the theoretical progress that helped physical chemists to lay a sound basis for our understanding of the properties of these phases, and hence for the reaction chemistry that took place in them, over the second half of the twentieth century.

12

Mg

Magnesium

Advancing the Technique

I F YOUR RESEARCH PROGRAMME IS BASED AROUND a particular technique or set of techniques then a part of what you do should be to advance the practice of that technique. Indeed, this is part of the established wisdom, developed by Professor Eric Von Hippel of MIT, of how innovations in such techniques occur, that is, by users of a technique as innovators. So a small stream of my research programme, both at Hunter College and later on, was devoted to that. Usually it was accidental, little digressions. Sometimes it was a substantive thing, like computerisation of the equipment at an early stage, before it was easy to do.

Because many of us were doing Raman spectroscopy with lasers for the first time, someone had the idea that we should have a little newsletter. Whether you had a problem, saw something unusual, helped to resolve a problem that others were having, whatever, you just sent in a letter to the editor of the newsletter, and every month or two you would get in the mail a copy of all the letters that had been received. In this way we became a community. And in science, as in other fields, community is very important.

In one of these newsletters two friends in the San Francisco area, Bob Snyder and Jim Scherer, each had a letter. Bob's was titled 'Wouldn't it be great if ...' and Jim's was 'It just so happens that ...' It is too difficult to explain the theory behind what Bob was suggesting, but it came down to if you could take the Raman spectrum of a liquid with two different positions of a polariser in the laser beam (something we routinely did), and then, using a computer, take one of these spectra, multiply the entire thing by 4/3, and subtract it from the other one, you could get some interesting structural information. Jim had a computerised instrument and did just that.

When I read this a little light went on for me. Going back to my undergraduate days, I had an interest in molecules that could be either planar or non-planar. My undergraduate problem was one where the non-planar version had a different symmetry to the planar. But there was a whole class of molecules where the non-planar version had no symmetry at all, and while this was a difficult structural problem to solve in liquids or solutions, by this computerised technique we could do it in minutes. An example would be the molecule shown below, chlorobiphenyl, in which all the atoms could lie in one plane, or the rings could be twisted with respect to one another. In the former case there is a plane of symmetry, and in the latter no symmetry at all.

And so we made a quick digression and did a number of these problems in a week. Now everyone could see that the technique had a power that was not previously known.

I also wondered if we could tackle such problems with solid powders. Normally this didn't work with Raman spectroscopy, because the polarisation data were lost by the light being scattered this way and that in the sample. Somehow lodged in my brain was the knowledge that when people made other microscopic measure-

ments on powders they used a set of
liquids, which you could buy, that
didn't dissolve the powder, but made it
seem like it was in a clear liquid. This
was accomplished by matching what is
called the refractive index of the liquid
to that of the powder. And with our
computerised instrument we could do

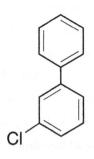

all sorts of tricks, like subtracting out the spectrum of the matching
liquid, then doing the 4/3 multiplication, etc. Another small
advance for the technique, and some more structural problems
open to quick elucidation. Recently my son Dave, who does research
in neuroscience, told me that refractive index matching liquids are
now being used to make brains appear more transparent.

Later on we coupled a gas chromatograph to our infrared spectrometer. Gas chromatographs separate volatile compounds, so if
you have a mixture they come out one at a time. Of course, what
you need to do is identify what is coming out. Today this is done by
one of a variety of so-called hyphenated techniques, such as coupling
the gas chromatograph to a mass spectrometer (gc-ms). Back then
this was not so common. We flowed the effluent from the gas chromatograph through a film of a cholesteric liquid crystal. These
materials, derivatives of cholesterol usually, have the property of
changing colour when even a trace of some organic compound hit
them. We hooked this up so that we could use the colour change to
trigger collection of the infrared spectrum, and that identified the
compound. As it turned out this was not very useful, other techniques were much better, but we at least had the satisfaction of one
of the reviewers for the paper in the journal *Analytical Chemistry*

saying that it would be among the most outstanding papers published in the journal that year.

I mentioned earlier that I got some grants just to buy instrumentation for my lab. One of these involved me at the forefront of a technique change as well. In 1969 I applied to the National Science Foundation for a grant, I think I asked for something like $35,000 to buy an instrument made by Perkin-Elmer Corp. to do far infrared spectroscopy, something very similar to the one I had used in Zurich a few years earlier. My proposal got very good reviews, and one day I was contacted by Fred Findeis, who was at NSF and responsible for such instrumentation grants. I knew Fred a bit, he had been a graduate student at Purdue some years before me, and we met at various conferences. On the phone he told me that he thought I had asked for the wrong instrument. He had visited a start-up company in Cambridge, MA, Digilab, which was making the first commercial Fourier Transform Infrared Spectrometers. This was not exactly a new concept, indeed I had heard a talk from Professor Gebbie of the National Physical Laboratory in England in 1962 doing this, but Gebbie had the resources of great machinists, computer programmers and electronics specialists behind him. Now, eight years later, it could be possible to get such an instrument commercially. This was in part due to the first minicomputers being available for the laboratory. Fred guessed that while Digilab had not yet delivered any of these instruments, they would cost about $50,000, so he offered to give me more money than I had asked for, if I was willing to take the risk to buy something not yet proven, but which would be the first instrument of the future rather than the last instrument of the past. Before the end of the phone call I had agreed.

At Digilab I already knew Peter Griffiths, who was to go on to have a fine academic career, and got to know all the other senior people really well. The instrument I wanted was not quite finished, indeed, when it was delivered to me it didn't really work, but together we made it work. Among the benefits of the Fourier Transform approach was that the minicomputer was integral to the instrument. One day I saw one of my first-year chemistry students carrying a load of computer manuals. I asked if he was studying computing, and he said no, he was the operator of the mainframe computer that the college owned, an IBM 360. I asked how much they paid him for that, and he said nothing at all, but it was good fun, and he learned from it. I had enough grant money to offer him a low-paid position, and together we integrated all the computer processing of the Raman spectra using the computer on the new instrument.

There were lots of other little technique things, then and later, and I loved doing them, and so did my students and co-workers. Science advances in this way, perhaps as much as by the big studies we do. Certainly the advance of chemistry from the 1960s onward was largely driven by development of half a dozen major techniques for elucidating problems of structure, bonding and reactions. After a time things that were difficult become routine, and older scientists look back fondly on the days when you really had to understand the details of how the instrument worked, how it was programmed, the electronics, etc. But that would have been the case a hundred years earlier with other techniques that we now routinely had done by someone else. There is too much to learn to have to know everything in depth, when you can trust instruments, or external labs, to do this for you. And maturity in science, as in other things, is about not being stuck in the past.

Awards

AWARDS TO INDIVIDUALS, AND SOMETIMES ALSO TO companies or universities, have become a staple of science. The Nobel Prizes more or less started this off, and now every technical society offers a plethora of awards. Some of these are named after individuals who made great contributions to the field, or subfield, and others are named for the donors. Sometimes there is a cash prize, sometimes just a plaque or a medal. Often in science there is an award address that the recipient gives, sometimes a symposium in his or her honour.

Even in my little field of molecular spectroscopy there are awards. One of these, the Coblentz Award, named after the American scientist W. W. Coblentz, who really founded the field of characterising molecules by their infrared spectrum, is for the outstanding young scientist in the field. Interestingly, the definition of young is subject to change. It was thirty-five when the award was created, then extended to thirty-six, and today it is forty. I won this award in 1975, and received it on my thirty-third birthday. Later on

I also received an award for my work in chemical education, the Oscar Foster Award, a distinguished research citation from Sigma Xi, the research honorary society, and a Gold Medal from the Society for Applied Spectroscopy. Some years later Purdue honoured me with a Distinguished Alumnus Award, which meant a lot to me.

It is wonderful to get awards for your research or for other efforts, like mine in trying to help high-school teachers do a better job of teaching chemistry. There is a feeling that your peers, or superiors, recognise what you have done and appreciate it. The people in the granting agencies who give money to academics also like it when you get an award for research they have funded, since it shows what good taste they have. But there is a difference between it being wonderful to get an award and it being important to your life as a scientist or educator, and I always tried to keep that perspective.

Things It Is Important to Know

O NE MIGHT THINK THAT A PRODUCT OF a systematic and generally high-quality education, such as I was, would have learned the key scientific and mathematical principles through the extensive coursework I had taken. To a great extent this was true, but over time I also found that there were several things that were important to being a functioning scientist, in all aspects. Functioning here doesn't just mean that you can do the right experiments in your lab, or know the theory behind them, though it certainly includes that. It also means that you can listen to a scientific talk and understand the key points even if you have never heard anything about the subject before. It means that when someone talks about a problem you can do some rough calculations in your head to approximate the answer. It means knowing if something being proposed is impossible because it violates basic laws. It means knowing certain trends or generalisations that hold true most if not all of the time. And it means knowing certain things in great depth so that you can begin to understand when the field is advancing.

For the science I did, and I believe for a lot of science, I found several things, some simple, some difficult, were crucial.

Too much education is about teaching students how to solve problems using particular formulae or equations. Students (and a lot of their teachers) think that these have to be memorised. Mostly, they don't, because if you simply think about the dimensions of the various terms in the equation you will find that you don't need to memorise it. Here is a simple example. A student wanting to know how far you travel in a certain amount of time depending on the speed memorises the formula

Rate × Time = Distance, r × t = d

but come the exam, he can't remember if it is r × t = d or r × d = t. However, if he knows that a rate is something like miles/hour, and a time is something like hours, then if you multiply a rate by a time, you also multiply the units, and if you multiply miles/hour × hour you get miles, which is a distance, because the hours cancel out:

mi/hr × hr = mi

We call this dimensional analysis, because instead of the actual units, like miles and hours, we can use what we call dimensions, distance, time, and know that speed is distance/time, and then work out what the equation must be to make the dimensions of every term come out correct. It is amazing how many complex formulae can be quickly constructed using this technique without need for memorising.

Related to this is approximation. Once you have figured out the route to doing a calculation through dimensional analysis, you can make some guesses as to how big the different terms are in the formula, try and get them correct to within what we call an order of magnitude, which is a factor of 10: for example, is the speed about a million miles per hour, rather than 100,000 or 10 million? Then you can do a rough calculation. When I am listening to a problem being presented, I often am doing these sorts of approximate calculations.

There is another aspect to approximation that I picked up early in my research career, and that relates to approximating the error in a measurement. We had a new device in the lab for measuring low light levels, called a photon counter. And I learned that the standard deviation, in lay terms something like the uncertainty, in a measurement by the photon counter was approximately the square root of the number it measured. Let's look at this in practice. Suppose the counter measured 100. The square root of 100 is 10, so we might say that the result is 100 ± 10, or a 10 per cent uncertainty. Now suppose instead that the signal is much bigger, so the counter measures 10,000 instead. The square root of 10,000 is 100, so 10,000 ± 100 which is only a 1 per cent uncertainty. The bigger the number, the smaller the percentage standard deviation. We see this in polling data for elections. If the sample size is 1,000 people polled, the square root of 1,000 is about 32, so about a 3 per cent uncertainty, which is a number you often hear quoted because that is frequently the size of sample that is polled. This is a very useful way of estimating standard deviation, and it stops you from getting seduced by the apparent accuracy of a number that someone gives.

These two techniques of dimensional analysis and approximation are so important that I wondered why no one ever taught them to me as an undergraduate. When Dave Beveridge and I were teaching first-year chemistry to undergraduates at Hunter College in the early 1970s, we met Professor Art Campbell of Harvey Mudd College, who had authored a radically different basic chemistry textbook that we liked (and the students mostly hated) in which he used these techniques to challenge students. We loved this, so we started to ask students to solve problems for which there was no formula, many of which Art Campbell had originated, like 'how many molecular layers are lost per revolution of an automobile tyre?', and the only way to do it is by making some guesses, and being sure you know how to go from thickness of tread back to molecular layers, via total miles travelled and revolutions to accomplish those miles. And no, you cannot do this by Googling it either. I hope it made them better able to learn later on, though I don't know if most of them ever figured out what we were trying to accomplish with our weird problems.

In chemistry, another one of the fundamentals is the periodic table. We have all seen it hanging on the wall of the room where chemistry is taught. But the arrangement of elements in the table tells us a lot about the chemistry. This is true both horizontally and vertically. For example, as we go across from iron to cobalt to nickel, which are adjacent to each other in one row as elements 26, 27 and 28, we expect a certain trend in properties (recall what I mentioned about the origins of my PhD thesis problem, where our research group was trying to study the properties of a series from nickel to cobalt to iron), whereas as we go down a column, say from nickel to palladium to platinum, we expect a different trend. I quickly learned how important it is to be able to think about such trends while listening to science being presented, even when there is no periodic table to hand. And the way to do that is to memorise the table. Every good chemist has pretty much memorised the periodic table, and understands the trends that it represents.

Then there is thermodynamics. Thermodynamics, and in particular the second law of thermodynamics, can tell you whether

THE SECOND LAW

The second law of thermodynamics works like this: there is a balance between heat content (we can think of this as the net energy after what is put in or given off in a chemical reaction, and order/disorder. The heat content is called enthalpy, and the order/disorder is called entropy. Now the balance between these depends on temperature, so if in a chemical or physical change (by a physical change I mean something like melting or boiling) there is a change in enthalpy, which we write as ΔH (Δ being the Greek letter *delta*, and the scientist's way of meaning 'change in'), that is heat content at the end minus heat content at the beginning, and the change in entropy which we write as ΔS, so entropy, or disorder, at the end minus entropy at the beginning, then a reaction or a physical change happens spontaneously when the quantity $(\Delta H - T\Delta S)$, where T is the temperature (and we use a kind of temperature scale called absolute temperature where T is always positive), is negative. Let's look at the example of melting. We have to put in heat to melt a solid and convert it to a liquid. So ΔH is positive, that is, the heat content of the liquid is higher than the heat content of the solid. But liquids are more disordered than solids. In a solid the molecules are in one place, whereas in the liquid they are moving around constantly. So when we melt a liquid ΔS is also positive, that is, the entropy of the liquid is greater than that of the solid. Now if both are positive, then at low temperatures the first term in the quantity,

a reaction is going to happen spontaneously or not. By this I mean, for example, if you mix hydrogen gas and oxygen gas, and send a little spark through the system to get it started, there will be an explosion and all the hydrogen and oxygen will have reacted to form water. But if you have some water, you can send a little spark through it from here to Doomsday and it will not spontaneously decompose to hydrogen and oxygen. You can decompose it to the ΔH term, will be greater than the $T\Delta S$ term, and the quantity will be positive. But as temperature is increased, there will come a point where the product $T\Delta S$ will be greater than ΔH, and the quantity will become negative. At that point melting will occur spontaneously, i.e. that is the melting point. Now let's apply this to a simple chemical reaction. When we mix hydrogen and oxygen and get the reaction started with a little spark they spontaneously react to form water, and a lot of heat is given off. We know the reason for this, namely, that bonds between like atoms (HH and OO) are weaker than bonds between unlike atoms (OH), so when you form the two stronger OH bonds you will expect heat to be released. That means that the ΔH term is negative for this reaction. What about ΔS? Well, the randomness of the reactants is greater than that of the product. Why? Because if you have a mixture of hydrogen and oxygen, if you could reach into the gas and pull out one molecule, you would only have a 50 per cent chance of knowing which it was, whereas in the product, pure water, you will always know. So the disorder of the reactants is greater than that of the products, hence ΔS is negative. Now our quantity is $\Delta H - T\Delta S$, so it looks like (-) – (+)(-). The second term will be positive at all temperatures, and the first negative. The overall quantity will be negative at lower temperatures, and positive at higher temperatures. Just the opposite of the melting example. But to know what temperature tips the balance requires knowing the size of each term.

hydrogen and oxygen, but you have to put a lot of energy in, either in the form of heat or electric current. The second law of thermodynamics will make it obvious that this is the case, and you don't have to know any numbers or complex formulae, you just have to know how this law works in practice. It is the second law of thermodynamics that explains why melting will happen at a definite temperature rather than gradually over a wide temperature range, so it was crucial both to our hypothesis on the formation of liquid crystals and to understanding why something different had been observed. Incidentally, Art Campbell's textbook, which I referred to above also emphasises this approach in teaching students of first-year chemistry, and I think it is the only such text to do so. The passion for the second law of thermodynamics, however, is very widespread among chemists, and when I came to live in the UK I got to know Professor Peter Atkins of Oxford University, who has written several books trying to explain what is happening in the world around us through thermodynamics, especially the second law. In the Preface to his book, *The Second Law*, Atkins wrote, 'No other part of science has contributed as much to the liberation of the human spirit as the Second Law of thermodynamics.' Sometimes I can hear a scientific opportunity being described in terms of 'We can do this and it will be the greatest thing ever', whether in a talk or in a business plan or grant proposal, and like any chemist worth his salt I am thinking, 'Does this violate the second law of thermodynamics?'

And finally, quantum mechanics. Quantum mechanics provides the main theoretical underpinning of all of modern molecular chemistry. It determines structure and to a great extent reactivity. All of the spectroscopic techniques I have referred to are predicted and explained by quantum mechanics. Now the problem here is that quantum mechanics is neither simple nor intuitively obvious. At least I found that I had to study it several times, then work at it, that is, use what I had studied, then study some more; only then was it deep 'in the muscle' so to speak, so that I could use it and understand its use without going back to books. In this sense it is like many things that are difficult to learn, be they physical skills or literary analysis. Malcolm Gladwell talks in his book *Outliers* about an apparent generalisation that those who develop proficiency in such things generally require about 10,000 hours. I have added up the time I spent learning and practising quantum mechanics and that is a good approximation. Within a factor of 10.

Polymers Everywhere: Tackling Ever More Complex Molecules, Their Structure and Reactivity, Becomes Possible

IN OCTOBER 1975 I MOVED FROM CITY University to Polytechnic Institute of New York. This was the same institution that had been called Brooklyn Polytechnic when it was my undergraduate home. Now I came, as a thirty-three-year-old, to be Dean of Arts and Sciences. Just about every faculty member was older than I was, and there were lots of difficult problems. Personally, the problem was one of gaining any respect from these senior faculty. There were two aspects to my life at Poly: one was learning basic leadership skills in a faculty environment where you have to lead through influence rather than authority, and the other was to continue my development as a scientist, because thirty-three is just too young to quit doing science. And only by doing both of these exceptionally well would I gain the respect I wanted.

One of my conditions for coming to Poly as dean was that I could continue to run an active research programme. A mentor of mine, Bryce Crawford of the University of Minnesota, had been Dean of Science there. He was a great spectroscopist and did research for many years as dean. And he put it to me this way: don't let 'deaning' get in the path of the beam.

To bring me, Poly agreed to buy me two instruments that I needed for my research: a new Fourier Transform infrared spectrometer and a new Raman spectrometer. Of course, I already had these in my City University lab, but technology had also moved on, so getting new instruments had some advantage. I had some leverage with the instrument companies, because I had just won the Coblentz Award and had an established reputation, so I was the kind of customer it was important for them to land. I negotiated both purchases very well.

In 1974, while on my way to lecture for three weeks in Bulgaria, I stopped in Paris, and made a side trip up to Lille, where great developments in Raman spectroscopy under the leadership of Professor Michael Delhaye were taking place. One little invention there struck me as interesting, among many. They had a simple, mechanical way of scanning a region of the Raman spectrum very quickly, say in two to four seconds. It occurred to me that there were processes that took place over a time scale of minutes, and if you could make a measurement every two seconds you could get information about these processes that was not available by other techniques. Some years after this, other developments would allow Raman spectra to be measured much faster, so that processes that took place in tiny fractions of a second could be elucidated, but I had some things in mind for the slower but still interesting processes. The Jobin-Yvon Corporation, with close links to the laboratory in

Lille, agreed to build one of the rapid scanning devices into the instrument I was to buy.

You will recall from my description of undergraduate research that Polytechnic had a long history of excellence in a field known as polymer science. This is the study of very large molecules, like poly-ethylene, polyesters, epoxy resins, polystyrene. And, of course, also biopolymers – proteins, carbohydrates, DNA. This excellence dated back to the arrival just after the Second World War of Professor Herman Mark from Austria, one of the greats in polymer science. When I was an undergraduate, Professor Mark had been my instructor in first-year chemistry, freshman chemistry as it was, and is, called. During my first semester as an undergraduate *The New Yorker* magazine published a two-part profile on Mark titled 'Polymers Everywhere'. Because Mark was there, others were attracted. While many of the most famous polymer chemists, such as Charlie Overberger and Murray Goodman, had since left, there were still at Poly such luminaries as Herbert Morawetz, Fred Eirich, and, more recently arrived, Eli Pearce. Because of its reputation, students were attracted to Poly to do graduate work in polymer chemistry, industry supported the department, and there was a great network of alumni. Moreover, and this was to prove important to my research efforts, senior people from foreign academic institutions and from industry either had some affiliation with the department, or came as visitors.

I had had a very good eight-year run on my studies of liquid crystals, but it seemed a good time to switch to something else, and the logical thing was to develop some new research strands associated with polymer science. At the same time I thought I could use my knowledge of materials like liquid crystals, and my skill and reputation in spectroscopy, to advantage in studying certain properties of polymers.

Crystallisation Kinetics of Polymers: Polyesters; Faster Techniques Allow Us to Watch Crystallisation as It Is Occurring

C RYSTALLISATION IS AN INTERESTING PROCESS. IF YOU have a concentrated solution of a solid in a solvent, water or some organic solvent, for example, it can sit there for a long time without any crystals appearing. Sometimes a seed is required, one small crystal, or even a speck of dust, and then crystallisation occurs around that seed. If the crystal grows very slowly, without dust or vibrations, it may be quite large, and 'perfect'. But if the process happens quickly there can be a huge number of tiny crystals, and even the small ones will have flaws. In this way we get very different-quality gemstones, or by the slow evaporation of water the clear sugar crystals of rock candy.

With polymers it is different. The molecules are so big that one can have regions which are crystalline and others which are more amorphous, not having order, gel-like even if they don't flow like liquids or liquid crystals (incidentally there are also liquid crystalline polymers, but while that was an active field of research at the time, leading to ultra-high strength materials like Kevlar, it was not the area in which I chose to work). And then polymers exist in so many forms – we can have thin films (like Cling Film or Saran Wrap), fibres, such as in our polyester or nylon clothing, sheets of thicker polymer, moulded forms for plastic parts or blown as in bottles for soft drinks. Each of these forms has important mechanical properties, sometimes also important optical properties (clear or opaque), barrier properties (bottles that do not let carbon dioxide, the bubbles, out), etc. And at least some of these properties are affected by the crystallinity, or the amorphous nature, of the polymer. These different forms – say film, fibre and bottle – can all be the exact same chemical composition, for example, the same polyester.

This was a new field for me, and with a few graduate students who joined me we set out to learn what was already known, both theory and experiment. There was a classic book entitled *Crystallization of Polymers* by Leo Mandelkern, and in it he had a long discussion of the rates of crystallisation, i.e. the kinetics, and how to describe these by a theory derived from a model.* Well, the first thing we did was to spend a few weeks going through this

---

* What does it mean to say, a 'theory derived from a model'? Well, a theory can be based on first principles, start with the basics of quantum mechanics, or for crystallisation possibly from classical mechanics, and perhaps some basic statistical considerations, and push forward to a mathematical description of the rate at which crystallisation will occur. This was then and remains beyond the scope of what is possible with quantum mechanical knowledge. So instead we describe the polymers in their amorphous state, then our 'model' says that a few crystalline seed regions will form, and around these crystalline regions will grow. Something will limit the rate at which they can grow, and something else will limit the size to which the crystals can grow. Probably each of these limits will have some temperature dependence that we can hypothesise. Then, with something like this model in our minds, we prepare a mathematical description of the rate (kinetics) of crystallisation. This mathematical description can then be tested against experiment, modified, and tested again.

material, re-deriving all the equations in the book relating to kinetics. This is important, because if you just read and make notes, without going through the intermediate steps, some of them omitted by the author, you will only have a superficial understanding, which will inevitably lead to mistakes when you try to apply the resulting equations to your own data. Well, interestingly, we could not get all the equations to come out as they were in the book. And after rejecting our own inadequacy as the cause, we concluded that there were probably a few mistakes in the text.

It was the days before email and instant contact. I could have written a letter to Mandelkern to ask him about our problems. But rather more fun was to invite him to Poly to give a seminar on his current research and, while there, to show him the problems we had found. This we did, and in the discussion, after looking at our results, he said, 'Well, I know there are a few mistakes, and the only way to check is to go to the Japanese edition of the book. We can't read the text, but the equations will all be easy to read.' It seems that when the Japanese decided to translate the book, they re-derived all the equations, and in this way found several mistakes. As it turns out, we had found some of these same errors. And so another good reason to treat even the contents of the leading text in the field with a bit of scepticism, but, more important, by working through this ourselves we deepened our understanding.

I wanted to work on polyesters, in particular poly (ethylene) terephthalate or PET, because of both its industrial importance and some fundamental problems that I thought we could resolve with our techniques. But I didn't really know enough about PET and its properties. Fortunately, Eli Pearce, my colleague and collaborator on other polymer problems I will describe shortly, invited Professor Menachem Lewin, director of the Israel Fibre Institute, to come as a regular visitor to Poly. Menachem and I discussed what I thought we could learn from the spectroscopy of PET, and he pointed me towards the right experiments to do. Moreover, it was fortunate that a big centre of expertise on polyesters existed nearby in New Jersey, at Celanese Corporation, and several of my friends, including a number who were in graduate school with me at Purdue, worked there. Through them we were able to get our hands on some specialised samples that were well characterised, and not readily available to others. Networks, collaborators and good students – the key to most successful research careers.

What we realised as we started to investigate the spectra of PET and related polymers (which were known as PPT and PBT, although in the UK they were known as 2GT, 3GT and 4GT, science having lots of nomenclature and shorthands to keep out the uninitiated) was that some of the literature was not quite correct. As a polymer was put under conditions where it could crystallise, it could also do other things. For example, there is a linkage in PET between rings which could look like either

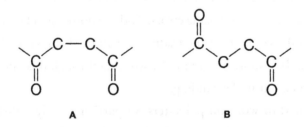

and sometimes the change from A to B was confused by interpreting data as showing an increase in crystallinity. We studied the whole crystallisation process, and as with previous work on liquid crystals we found that it took place in several steps. If you heat the polymer there is a long period where nothing happens. Then, presumably once nuclei have formed, there is a rapid crystallisation stage, after which the process comes to a nearly complete stop. There must be some barrier, and after this is overcome, the crystallisation process resumes. This complex behaviour had not been seen before, because there hadn't been a probe of the kinetics at the molecular level. Then we were able to probe the vibrations of the crystalline lattice directly (but only because we had computerised instruments that could perform some arithmetic manipulation of data between samples in different states) and still more details of the crystallisation process emerged.

17
Cl
Chlorine

Stale Bread and Crispy Cookies: And We Can Apply This to the Complex Problem of Food Spoilage

ONE OF THE THINGS MY STUDENTS AND I came across while reading about polymer crystallisation kinetics was the idea that when bread or cake becomes stale the main thing that happens is a crystallisation of starch molecules. Starch is made up of two polymers, amylose and amylopectin, both polymeric forms of the common sugar glucose. Amylose is linear, that is, one glucose molecule after another in a long chain, while amylopectin is branched, that is, every now and then the glucose attaches differently so that a side chain forms. While the macro properties of polymers don't always reflect the micro, in this case they do, so when you have a sticky starch, as in certain kinds of rice, for example, it is high in amylopectin.

Now when you bake bread, many things happen. But the one most interesting from the viewpoint of staling is that the starch, which you have mixed with water in forming the dough, goes from a more crystalline state to an amorphous state. It is this amorphous form which gives bread the desirable texture. At room temperature, the most stable form of the starch-water gel is a crystalline form, and that is what we would associate with stale bread. Over time the starch-water gel reverts from the amorphous to the crystalline form. Why, if the crystalline form is more stable, does it not revert immediately? Well, if you have ever baked a loaf of bread, or a cake, you will always see the instruction 'When the bread has been baked, remove from the oven immediately, turn out of the pan, and let it cool on a rack.' By following this instruction, what you are doing is quenching the gel in its amorphous form, not giving it time to crystallise; you are essentially keeping it trapped in an unstable form. Of course, there are little seed crystals present, and over time they will grow into longer crystals and this is what we know as stale bread. If you were to turn off the oven and leave the bread in there, letting it cool very slowly, then you would find that you had baked a loaf of stale bread. It's that simple. Incidentally, many people think that staling is just the loss of water from bread, the same as drying out. We know from experience this is not the case. If you take bread that it slightly stale and put in the microwave for thirty seconds you can restore the fresh texture, and without adding any water. But if you do this and don't eat it quickly, then it will revert to its stale state very quickly. What are you doing in this experiment? The starch has started to crystallise, but by putting in energy in the form of microwaves, you heat it back to the gel state. However, the nuclei around which crystallisation occurs are mostly still there, and so the starch will rapidly crystallise again.

All of this was known at the time, but very little was known about the details of the staling process (known to starch chemists as the retrogradation of starch), that is, what happens at the molecular level. I had filed away this interesting subject in some corner of my brain when, one day, a colleague at Poly, William Winter, who worked on chemistry of sugars, asked if I would meet with Dr Iain Dea of Unilever Corporation, who was over from England and wanted to talk to me. Of course I did.

Iain was an accomplished researcher in Unilever's labs, and our work on crystallisation kinetics of polymers had caught his attention, and that of his management. He asked if he could come to spend a year in my lab, paid for by Unilever, to learn the techniques we were using. We chatted about this, and I had only one request of him: Our research was based around two main instrumental techniques, and what limited our ability to produce results was time on these instruments. He could come to the lab, but he had to work on problems of interest to us, because I wanted to maintain the focus we had on two major problem areas. He agreed, willingly, but in the course of subsequent conversation I mentioned the problem of bread staling and we talked about it a bit. Then he said, 'Now, if you were willing to work on that, Unilever would be really interested in my coming here.' So we agreed. Once again, as with Menachem Lewin, I had found a great senior collaborator.

Some fields of research are fairly open, by which I mean that new entrants are always coming in, there are many groups around the world and a large, ever-changing community exists. Spectroscopy, my core discipline, was like that. But starch was at the other extreme. There were a few key laboratories, most of those working in the field had trained at one of those labs, and they all knew and recognised each other. So when someone new, like me, comes along with some results, using a technique none of them have, and giving information about a key process that was previously inaccessible, it gets their attention and possibly creates a little resentment.

But you don't get to results that can be presented at once; first there is work to be done. Iain and I started to do some experiments, but I realised that if I was to have this as a subtheme of my research on polymer crystallisation kinetics I needed some additional funding, to pay for at least a graduate student or two. Unilever might have given this, probably should have, but as it happens several of the companies in this area of research had set up a basic research fund through something called the Corn Refiners Association, and I was advised that we could get some funding through that. So once we had a few quick preliminary results, Iain and I made a trip out to Illinois to a company called A. E. Staley, where we could present our case, and talk about such interesting topics as to how to decide the stretchiness of a croissant, and then to New Jersey to National Starch. With Unilever on board as well, we now had sufficient support. Of course, for other basic research one would go to a Federal Government agency, National Science Foundation or the like, and in principle there should be someone there who appreciates that staling is one of the most important worldwide problems, involving as it does massive food waste. My guess was that such a person did not exist. Another possibility was that when a proposal related to starch came in to NSF it would immediately be dispatched to one of the core group of starch researchers, who, seeing this was from an 'outsider', would find a way to dump all over it. Such is the nature of scientific funding sometimes. There were lots of other reasons not to go that route, most obvious of which was that with the Corn Refiners Association I could write my idea down on five pages or less, spend a day each in Illinois and New Jersey, and have a very high probability of success very quickly. With the support-building work done, we put in a proposal and immediately got the funding we needed.

We were not the very first to look at the Raman spectrum of the polymers amylose and amylopectin, which compose starch. My friend and colleague at Case Western Reserve University, Jack Koenig, had looked at these polymers some time earlier and done a bit of interpretation of what he had seen. We were able to build on his work and greatly extend it. Iain and I looked at a whole variety of starch-water gels, with different starches, everything from the pure polymers to corn, potato, waxy starches, etc. We identified which bands in our observations were sensitive to addition of water and compared spectra at high and low temperatures. We had to do a lot of interpretation of the basic data, and, having put forward a hypothesis for each, used a number of model systems to confirm or refute our hypothesis. Then, using our rapid scanning technique, we took spectra every thirty seconds over many days, as we watched the staling process in detail, at the level of the individual chains in their environment. Fortunately for us the process turned out not to be a simple one. Rather there was an initial step that took you part way there, and then several distinct later steps. This was interesting, because it offered the possibility of intervention at different stages, to achieve particular textures.

As I said, only a few groups studied this, and when we appeared at conferences and started to show our results, we got attention. One day my phone rang, and the caller identified himself as an executive of Proctor and Gamble. 'We understand you have been working on retrogradation of starch,' he opened. 'Yes,' I replied. 'I wonder if you have heard of the chocolate chip cookie war?' Hmm, this could be fun, I thought. It turned out that the ideal chocolate chip cookie is crispy on the outside, soft and chewy on the inside. Crispy is the equivalent of crystalline/stale if you will, and soft and chewy is gel-like. Fine if it is fresh from the oven, but how could this possibly be maintained in cookies that were mass-produced, packaged and sat on supermarket shelves? A competitor of Proctor and Gamble had managed to achieve just that. Could we help? Actually we were able to do a few experiments to show how this was done, and it was simpler than even they imagined. If you spray sugar onto the baked cookie so that it penetrates just a millimetre or so in, it induces a rapid crystallisation, which stops at the barrier where the sugar concentration falls to zero. The crystallisation, because it is around the sugar nuclei, does not spread.

About a month later, perhaps after another paper was published from our work, the phone rang again, this time a caller identifying himself as from the Nabisco Corporation. 'We make a product you may have heard of. It is called Shredded Wheat.' Yes, I had heard of Shredded Wheat, after all I am an American boy and loved to eat it as a child even though it didn't have much taste, because there were cards in the boxes with pictures of famous cowboys. So, what was his problem? Certainly not the Shredded Wheat war. 'Well,' he said, 'Shredded Wheat has a unique texture, and after we acquired this cereal back in 1928 from the inventor, we thought we could make Shredded Bran, Shredded Corn, Shredded Rice. But no, we were never able to duplicate the texture, and it even requires a specific kind of wheat. Despite decades of work, we have never been able to figure out just what it is about this wheat which gives the Shredded Wheat texture. Of course, today, with genetic modification, if we knew what property gave the texture we could perhaps engineer it into other grains. Might your techniques help?'

Well, this was a more difficult problem, and I was not sure we wanted to tackle it, but I told him I would give him a price to undertake a two-year study of the problem, which I did. For some reason they never funded the work, we didn't look at it any further, and I think the only advance to this day is that they have put raisins into the Shredded Wheat.

Anyway, the work was a lot of fun, and when the university PR department heard about it, they contacted a New York newspaper, *Newsday*, who were sufficiently intrigued to do a column on me titled 'Fresh Baked Polymers'.*

---

* My work on polymer crystallisation kinetics was carried out by graduate students Frank di Blase, Marianne McKelvy, Ellen Miseo, Jim Sloan and Y. Kwak.

Polymer Degradation Processes: Probing Chemical Reactions in Complex Molecules Exposed to Heat, Light and Other Extreme Conditions

PROFESSOR ELI PEARCE HAD COME TO POLY from the Research Triangle Institute in North Carolina, a few years before I arrived, having been a graduate student at Poly with Charles Overberger many years earlier. Eli was Chairman of the Chemistry Department, and I was Dean of Arts and Sciences, so his boss in the hierarchy, but given that this is an academic setting, there was not that much bossing to do. And we were also colleagues in the Chemistry Department, where I held the rank of professor in addition to being dean. Eli was interested in what happens to polymers as they degrade, whether due to heat, oxygen or light. Some polymers are subjected to extreme conditions, for example, the epoxy resins that were used to attach the ceramic tiles to the surface of the space shuttle, and NASA was funding Eli's work for just this reason.

Eli knew a lot about the chemistry that could occur when polymers degraded, and I knew how to determine what was happening from the infrared spectra. So we were a perfect team. Moreover, our combined reputations made it relatively easy (because it is never easy easy) to get funding for the work. We jointly supervised a large group of graduate students and postdoctoral fellows working on these problems over about eight years.

On the one hand, my role in this was fairly straightforward. Eli and his students would suggest that one of two or three things could happen, and I would help them interpret the data to sort out which was correct. As usual, we benefited from the students hearing each other speak of their experiences and frustrations, so once a week, for several years, I gathered the group in my office. This started as a necessary exercise to allocate time on the various instruments for measurements. But to make that allocation I had the students explain what they had done, where they had progressed to and what had to happen next. I think everyone learned from these little expositions each week. And because all of the students were from Asia, it also gave them practice in explaining their work in English, practice very much needed, and in a non-threatening setting.

One week a student showed us a new approach he had developed to treating the data, so that he could tell which groups in the polymers were more stable and which less stable. I thought it was a brilliant idea and asked him how he thought of it. He said, 'Well, Professor, you told me to do this last week.' In fact I had told him something completely different, but he misinterpreted what I had said and developed this original idea. It became one of the core techniques we used for several years thereafter.[*]

---

[*] The work on polymer degradation using infrared spectroscopy was carried out by S. C. Lin, Mo Yeen Ng, C. S. Chen, C. H. Do, M. S. Lin and S. P. Ting.

19

K

Potassium

Some Thoughts on Polymer Science

IN ANY SCIENCE, THERE ARE MORE AND less prestigious areas in which to work. Of course, these can change over time, but there is a snobbishness to science that cannot be denied. When I was an undergraduate we took three courses in aspects of analytical chemistry, and our professor, Lou Meites, said to us somewhat wistfully that none of us would become analytical chemists, because as a somewhat elite bunch we would be pushed into the more prestigious areas of organic, physical, biochemistry or possibly inorganic chemistry. This was true for those who went towards academic careers, but for some of my classmates who headed to industry analytical chemistry was part of a big job market.

Now polymer chemistry was interesting in this regard. At Brooklyn Polytech, where I was an undergraduate, there was a big contingent of polymer scientists, and it was the thing the chemistry department was best known for. A few other universities also developed strength in polymer chemistry, including Case Western

Reserve, Akron and the University of Massachusetts. But when I headed to graduate school at Purdue, there were no polymer chemists at all, and that was pretty much the case for the chemistry faculties of all the top twenty graduate schools. Imagine – the part of chemistry that was the basis for most of the multi-billion-dollar chemical industry, polyethylene, nylon, polyesters, etc., and none of the leading academic centres in the US thought it worth having a single faculty member doing research in this area. During my four years at Purdue, where we had at least one, more often two, visitors giving seminars every week, I cannot recall a single seminar on polymer science. It was clearly too dirty to be worthy of the time of faculty and students.

This was much less the case in Germany, where from the early part of the twentieth century polymers had been a topic of academic research and furious arguments. The polymers known and characterised in the early decades were all biopolymers, things like cellulosic polymers, such as starch, cellulose, rayon and cotton, proteins including silk and others of biological importance, etc. Because much of this characterisation yielded a variety of results for properties such as molecular weight, some argued that the polymers, or macromolecules as some called them, were not actual molecules at all but aggregations of smaller molecules. This controversy involved people like Wilhem Ostwald's son Wolfgang, Hermann Staudinger, Herman Mark and others. In the 1930s it inevitably degenerated into accusations of Jewish scientists attacked by 'Aryan' scientists. In the end it became clear, but only in the late 1930s, that the macromolecules were indeed bound together in exactly the same way as smaller molecules, but that they could be shorter or longer in chain length, linear or branched, crystalline or amorphous. And that all of this variability opened up an array of possible applications.

While some synthetic polymers had been produced in the nineteenth century, this was largely by accident – leaving small molecules like styrene or vinyl chloride exposed to air – and they were thought to be oxidation products. The great period of polymer synthesis was the mid-1920s to the 1960s, with polyvinylchloride at Goodrich in 1926, polyacrylonitrile at IG Farben in 1930, polyethylene at ICI in 1933, Nylon in 1935 and Teflon in 1938 at DuPont, polyethylene terephthalate (Mylar, Dacron and other names) at the Calico Printers Association in Manchester, England, in 1941, Orlon at DuPont in 1946, polypropylene at Philips Petroleum in 1951, and then Kevlar and Nomex at DuPont in the 1960s. Important work was done at Montecatini in Italy by Giulio Natta and Karl Ziegler at the Max Planck Institute for Coal Research in Germany, especially on catalysts using the newly characterised metallocene molecules (sandwich compounds where a metal is surrounded by two organic rings) to produce polymers of a particular 'tacticity', that is, where certain groups, for example a repeating CH_3 (methyl) group or a chlorine atom, are always on the same side of the polymer chain (isotactic), on alternating sides (syndiotactic) or random sides (atactic), which could lead to very different properties and applications for the same polymer.

Even in the late 1960s the conversation in polymer science was that there were few new polymers to be made that would achieve the same scale in application as these and others that were synthesised during the second and early third quarters of the century. This has proved to be the case, although there have been great advances in catalysts used to produce these polymers, control of properties such as molecular weight, understanding of crystallinity and many other properties. Again, this understanding largely arose through a combination of spectroscopic techniques, theoretical studies (some by physicists rather than chemists) and organic synthesis. Gradually in the later part of the twentieth century a few leading chemists and physicists began to study polymers at leading academic institutions – Paul Flory at Stanford, Pierre-Gilles de Gennes in Paris – and a number of other leading universities saw fit to allow this once dirty area onto their faculties. MIT, for example, has had a substantial effort in polymer science in its chemistry department for some years, as has Caltech.

20

Ca

Calcium

Transition: A Scientific Life in Academia Exchanged for One in Industry

I CAME TO POLY IN 1975, AFTER EIGHT years at Hunter College. I had many roles at Poly, Dean of Arts and Sciences, Professor of Chemistry, then Vice President for Research and Graduate Affairs, and concurrently Director of the Institute of Imaging Sciences. I was busy and challenged and got to do things I had never thought about doing, some of them outside of my personal laboratory science. Sometimes I just applied a little systematic thinking to a problem. For example, as Vice President for Research and Graduate Affairs, I was responsible for the income we achieved through tuition received from graduate students, particularly the part-time students who came to Poly in the evening, usually sponsored by their companies, to get a master's degree or learn some new techniques. I asked our Vice President for Finance for some money for an advertising budget, got an agency to help and started advertising. We put in place some ways of meticulously tracking responses to the ads so we could do experiments and assess their effectiveness – bigger vs. smaller ads, coupons clipped and mailed in vs. phone calls, whether a toll-free number helps very much, Saturday or Sunday better – and monitor the payback on the investment. Seems straightforward? Yes, but I learned from *The New York Times* education people, who found out about our methods, that no one else was doing this. They had me come in and show them our methodology. This was my scientific training being applied to a new problem.

Poly gave me a chance to learn about leadership, to test and greatly expand my personal skills and competencies. Sure, at Hunter College I had been elected as Chairman of the Chemistry Department, and I had two interesting and very different role models for leadership there, F. Joachim Weyl, who was our Dean of Science, and Jacqueline Wexler, who was President of the college. But the range of problems I faced was pretty modest, and unless you have to deal with a great variety of challenging issues you can't really develop as a leader. At Poly the range was greater, and I had both good and bad role models to observe and learn from.

I became Director of the Institute of Imaging Sciences (IIS), a bit by accident, as the first director resigned suddenly. The Institute had been founded following a million-dollar gift from a private donor, Greg Halpern. Its activities spanned chemistry, physics, computer science and electrical engineering. In a sense it was far from my areas of interest, though my scientific background was broad enough so that I could understand the technical challenges of the industry pretty quickly. Soon I was an invited speaker at conferences around the world on imaging science. And I was able to speak about it coherently enough to lure about twenty industrial sponsors to the Institute. I learned a lot from the executives who represented these industrial sponsors on our advisory board, from diverse companies like Kodak, Xerox, Fuji, Fairchild and others. I brought in Dr Arnost Reiser, who had recently taken early retirement from Kodak in England, as a deputy director, and he added a great deal of scientific strength to our efforts. We had all sorts of interesting seminars and a lively newsletter, and of course I used some of the money to convince faculty to work on things of interest to the donor and the industrial sponsors.

One of the incidental results of my role at IIS was that I met a Poly alumnus who was CEO of a company called Optronics International. He was entitled to appoint one member to the board, and asked me to be that person. This was a troubled company, with a good product, and I learned a lot. My fellow board members included a Boston-based investor, the head of a German company who was a customer as well as an investor and a leading physician who used imaging techniques in his medical practice. This was really the beginning of my immersion in a business, learning how to look at the numbers, when to challenge and when to praise, how pressure could be used to drive performance, and when it was ineffective or had the opposite effect. Yes, I contributed from my scientific knowledge, and my leadership skills, though still fledgling, were better than those of some in the room, but I was also a student there. Whatever we do, whatever we are paid or volunteer to do, should be a mix of contribution and learning throughout our lives.

Not long after I joined the board of Optronics, a friend from undergraduate days, Jud Flato, was involved in the buyout of a company called Spex Industries, which made instruments I had used in my lab. Jud had been a scientific leader at a company called Princeton Applied Research, and while he was there I had consulted on a project of theirs, so when he contacted me again and offered me a spot on the board of Spex I was enthusiastic about accepting. In contrast to Optronics, now my technical expertise was directly relevant, and I even knew a little about how to assess the business performance. Spex was backed by some interesting venture and finance guys, and for a time we had a good ride with the company. Later there were problems, as there were with Optronics; in both cases the solution was to sell the company, and participating in the decisions around those two sales was also very instructive.

So there was a period, from about 1982 onwards, when I felt I was transitioning away from my academic role. Part of this was that I was learning about business, and finding it more interesting than I had expected it to be. Part was that I had a certain restlessness to do something different. I know that many people get a PhD, do a postdoctoral year or two and then take an academic post, after which they are happy, indeed feel privileged, to spend an entire career doing research, teaching, publishing papers, maybe a little consulting. I realised I was not one of them.

Deans frequently move on to be university presidents, and I was nominated and shortlisted for two college presidencies in the early 1980s. I was still only forty years old, and I am sure if I had stuck at it I would have got one sooner or later. But something else happened, somewhat unexpectedly, which took my scientific life off in another direction entirely.

One of the first people I met as I was becoming a spectroscopist was Jenny Grasselli. In 1962, just before I started graduate school, I went to a meeting on Far Infrared Spectroscopy in Cincinnati, and Bob Bauman introduced me to Jenny. Over the years we saw each other frequently at conferences, and when I was a faculty member most often at something called the Pittsburgh Conference, held each year in March, usually not in Pittsburgh! Jenny and I were part of a small group of friends who usually reserved one evening at the conference for dinner, sometimes even for dancing at a local nightclub.

Sometime in the late 1970s Jenny asked me to collaborate on an article for a physics review journal on how chemists used Raman spectroscopy. When we finished the article it was pretty long, and became the entire issue of the journal. Fortunately, we realised that it was actually long enough to be a book, and the journal agreed that we could retain the rights to publish it as a book. We sent the manuscript to John Wiley and Sons, the leading publisher of such technical books, and they agreed that if we put in an introductory chapter, suited for a book but not the expert readership of a journal, they would publish it. Thus our book, *Chemical Applications of Raman Spectroscopy*, on which Jenny's associate Marcia Snavely was a co-author, appeared in 1981.

By 1983 Jenny was one of a trio of senior managers running the R and D laboratories of Standard Oil of Ohio, in Cleveland, reporting to Glenn Brown, who was on the board of Sohio. One day, in March 1984, at the Pittsburgh Conference in Atlantic City, Jenny asked to have a little talk over lunch, and in the course of that talk she suggested to me that Sohio was looking to expand its R and D laboratories, that they could of course hire young people, but that they had a gap in leadership and scientific positions for people in their forties. Given my academic management roles and my scientific reputation, they were interested in my joining them. Would I consider it?

Now I had never worked in industry for a day in my life. Not even a summer job. I had no idea what it was like, and whether it suited me. But I was restless, and I realised that part of this restlessness had to do with the scientific research part of my life, not just the ambition I have described. There were two parts to this. The first was the eternal need to get funding. I wrote proposals, mostly to the Federal Government, like everyone else; they generally got very good reviews, and some of them got funded. But even when I was successful the amount of money was not really adequate to do the research I wanted to do. I know that later on the National Science Foundation and other agencies came to appreciate this problem of grants being too small, but of course the only way they could solve it was by reducing the success rate. The other problem was that while I had lots of students, and some of them very good, the work that I wanted to do was quite challenging mathematically, and most of my students were not up to it. So I knew that I was suiting the research I did to the quality of the students and the funding that was available. This is not satisfying. I wondered whether, in a management role in an industrial research lab, I could still have a scientific life, and further, whether the funding, and the colleagues in the lab, would solve both of these problems.

My interview in Cleveland was very interesting. At dinner with the three lab directors, they told me that Sohio had a acquired a lot of businesses, mostly as a consequence of buying a conglomerate, Kennecott, at the heart of which was a copper mining business, but appended to it were many other companies. The job of R and D as these guys saw it was to produce technology that could make these businesses really profitable, thus building the foundation for a post-oil company. They also had set up a number of new ventures completely disconnected from the oil business. In their minds, the exploration business more or less took care of its own technology, and the refining/marketing business, which was a big part of Sohio, was uninteresting from a technology point of view.

I didn't say anything, but I did start then to think about this.

The second crucial meeting I had was with Glenn Brown, who was the managing director responsible for technology. He told me that if the company was going to hire a senior person like me they didn't want to make a mistake, and they didn't want me to make a mistake. So he thought we should both take time to explore the idea, to get used to it. This was very unlike most industrial hiring, where there was an interview, an offer and pressure to come to a quick decision. But Glenn was, and is, a smart guy.

So we worked out an approach. I would become a consultant to the company. Jenny would select some of the regular reports being written by the R and D centre staff, and I would be sent these. I could comment, ask questions and raise technical or business issues. Then, about once a month, I would visit and meet with people. So I would get an idea of what was being done, why it was being done and who was doing it. I thought this was a good way to address some of my concerns about joining them, or any industrial research lab, and for them to address their concerns about me. Glenn thought we could take a year to do this.

Meanwhile, another concern had to be addressed. I had a wife, Susan Lees, who had her own career as a professor of anthropology in New York, had no desire to move to Cleveland and probably a limited array of possible jobs if she did move. But Sohio were prepared to work this problem with us as well, and they did. As often happens, it was not a very satisfactory solution, because she liked the job she had in New York City and she liked New York better than Cleveland. But it was enough to get us to agree to the change, one of us reluctantly.

I spent about thirteen months consulting with Sohio Research, talking to various staff, visiting the lab and getting to know a few colleagues really well, some of whom remain friends many years later. I found it stimulating, and of course they paid me for it as well. In the end I did get an offer, and after some hesitation and tinkering with the terms, I accepted. After that my entire life changed, including my scientific life.

The Reproducibility of Results:
Laboratory Robotics Change What
We Can Measure and How We Can
Do It Better Than with People

Y FIRST JOB AT SOHIO, DESIGNED TO introduce me to the company and vice versa, was as Director of Analytical and Environmental Sciences (so despite the prediction of my undergraduate teacher, Lou Meites, I was managing one of the world's greatest departments of analytical chemistry). It was a substantial management job, not just in terms of the number of people and the budget, but because the laboratories for which I was responsible did work across all of the business of the company, both the established and start-up businesses. Almost all the techniques we used were very familiar to me, I had either worked on them myself or understood them well. But the problems being tackled were all new, and I had to learn, and learn quickly, what was really of importance, and how we could make a difference. In these cases, making a difference did not necessarily mean advancing the science, it usually meant helping the company be more profitable.

These problems were very diverse. Some of them involved the most basic things in chemical analysis, such as a laboratory at a refinery where it was necessary to determine, several times a day, the strength of a sulfuric acid sample drawn from a reactor on site, and this was being done using titration with a phenolphthalein indicator. I had done this experiment as a high-school chemistry student. Now the challenge was: could we do it without human intervention, to an accuracy and precision ten times better than what was expected in high school? The great advantage here of a central research lab, keeping up to date on the latest instrumental techniques, was clear. The supervisor of a refinery lab cannot be expected to take the time to stay current, let alone evaluate which of a number of new techniques, and from which manufacturer, could meet his needs. But the central group that I headed was expected to do this and provide the solution immediately, even before the refinery knew there was one!

Another part of the company was developing a gold mine in Papua New Guinea, on Lihir Island. The gold there is abundant, at a very high concentration in soil, but so finely divided that it cannot be seen by eye or conventional microscopes. Could we develop a technique for imaging the gold in the soil? In this case, it was not the most basic chemical instrumentation but the most advanced that was required for the job. We had foreseen the need for this, hired a young PhD the previous year who knew the technique and purchased a $150,000 instrument for him to use. Within months of his arrival the invisible gold problem landed on our doorstep, and we were able to repay our investment many times over.

On the north slope of Alaska, in the huge Prudhoe Bay oilfield, there was a dispute between the shareholders about ownership of the reserves, which hinged on analytical questions about the composition of the oil, gas and lighter hydrocarbons present. Each oil company had its own team trying to prove its position, and the amount of money at stake was huge – billions of dollars. What this means is that a company can 'find' more oil by winning such a dispute than it can by exploration over several years. We did our own analysis to back up the Sohio case, but in the end our results were not the definitive ones. I was given to understand that our analysts were as good as the very best in the world at this, but Mobil and Exxon (two separate companies at that time) proved to be better. However, none of the results were definitive, and the dispute rumbled on for years, finally being settled by negotiation rather than science.

I could go on listing hundreds of problems of this sort. It was one of the most enjoyable periods of my scientific life, both during the period of my consulting before I joined the company, and in the first eighteen months afterwards, when I got to learn about these problems and apply my scientific brain to moving them forward. As important, I was for the first time, in a substantive way, immersed in the intersection between science and business, a space that was to occupy me for the remainder of my scientific life.

But I also wanted something, even a small something that I could work on myself. As was the case ten years earlier when I became a university dean, I still was not prepared to give up on doing some science myself. In the lab, I had a brilliant Russian émigré scientist, Mike Markelov, who was working on the use of robotics in the lab. Mike's original motivations for using robots to replace human analysts were throughput (robots don't take lunch breaks) and hazardous materials that were dangerous for human analysts to handle. My interest was in whether they produced much more reproducible results, day in and day out.

In the process of examining this, I looked back at some classical work that I had studied as a first-year graduate student on how errors made in a measurement propagated through the calculations. If I wanted to understand reproducibility of results I needed to know what happened to the variability in measurement as the data became a computed result. Think of it this way: if I am trying to determine the amount of benzene in a sample of gasoline, I measure the size of a peak on some instrument, and the volume of the sample. Now to get the per cent benzene, I have to do some subtraction, division and maybe a square root. If there is a 1 per cent uncertainty in each of my measurements, what is the uncertainty in the final per cent benzene? It is definitely greater than 1 per cent.

I also found that while any given lab would have its own procedures for carrying out a series of measurements on a sample, another lab, charged with the same task, would do things slightly differently. The second laboratory analyst, even following the same written procedure using the same equipment, might consciously or unconsciously introduce either a biased result or one that had different degree of precision. Analysts might produce different results on the same sample before and after lunch, on Monday morning versus Friday afternoon. The variability of results between labs, especially when a small difference at the measurement end could lead to a larger difference in the final result, meant that inter-lab comparability was often very poor. Indeed, at the time I was doing the work, doping of Olympic athletes was in the news, and I realised that while they required a result to be reproduced with two different samples, the so-called A and B samples, these were always analysed in the same lab, usually by the same chemist. If they had sent the A and B samples to two different labs the chances of taking away a gold medal from someone would be much smaller.

Robots went a long way to fixing this problem. They did every task exactly the same way, and took the same amount of time to do it. Too often the time taken by different analysts is a variable that is ignored, causing a lack of precision in the results. This was something I already knew from a part of my PhD thesis research. The robots were so beautiful to watch in action too. Mike had several films made of our robots at work, with dramatic background music, and we presented this work, including the films, at several conferences.

Mike was also keen to pioneer switching some of the analyses that were done in the refinery and chemical plant laboratories to online instrumentation. In a refinery or chemical plant one needs to check frequently that the process is running as it should, that no one has dropped their lunch into the reaction vessel or added manganese when they should have added magnesium (both of these actually have happened in my experience). So the standard procedure was for a technician to go out with a sample vial of some sort, open a tap built into the plant especially for this purpose, let some of the material run out (usually onto the ground, but later we stopped them doing that) and then take a sample back to the lab for analysis. Of course, by the time it had been analysed, and a problem detected, a lot of bad product would have been made. This was what we called offline analysis, and now we wanted to switch to online, where the instrument to do the analysis was built into the pipe or reaction vessel.

Some simple stuff of this sort had already been done, such as pH measurement, but now ruggedised instruments for more complex measurement were appearing in the marketplace that could withstand the rigours of the plant environment. Recognise that up until then scientific instrumentation was made for an air-conditioned laboratory and designed to be used occasionally during the day. If we want to put an instrument into a pipe at the Toledo Ohio refinery, it must withstand temperatures from -20 to +100°F, rain and hailstorms, and the abuse of field operators rather than lab technicians. Or, as an alternative, the instrument is in a building, and there is a fibre optic probe in the pipe with a long cable to the place where measurement is done.

In the chemical plant at Lima, Ohio Sohio was making acrylonitrile, a starting material for many plastics, and as a by-product HCN, cyanide to you. And the purity of the HCN was a key challenge, because the main customer, DuPont, specified the minimum purity that they would accept. It had to be measured several times a day, and this meant an operator going to the tank where it was stored after production, withdrawing a sample and taking that to the lab for measurement. Considering the very real danger in working with these cyanide samples, this seemed a good target for us to start with for online measurement.

Mike did several tests to prove the feasibility of this, and then he and I went to the chemical plant and presented a proposal to the assembled management. There were lots of questions, and we could deal with all or most of them. But there was also a certain hostility to guys coming from the central lab and showing them how to do things better. So at some point, a senior manager expressed great scepticism, and this was all that was required for several of his juniors to pile in as sceptics themselves. Fine, I said, if you don't want it, you don't want it. I gathered our papers together and told Mike we were leaving. Immediately they all backtracked and insisted we stay and discuss it more.

So we did, but there was no action. This is the industry version of political correctness: agree to something that senior management seems to want, but do nothing to implement it. Then, by chance, we had our opportunity. There was a strike of the employees of the plant, and, in preparation for this, staff at the laboratory were trained to operate the plant during the strike. Not just the lab of course. Head office staff worked in the plant as well, even running the cafeteria. Mike joined those who were being trained, and naturally enough got himself assigned to the HCN plant. The strike lasted two weeks, and that was all the time that was necessary for Mike to automate all the analyses for which he could bring equipment to the plant at the start of the strike. By the end, instead of having a full-time job running tests, Mike could sit back and let the instruments control the process.

We had a big budget for new instrumentation in our laboratory, and some of these were major purchases. For many years there was a budget, and specific items were included for each year. It was a generous budget, and I think generally well spent. You can imagine that we got a lot of attention from the major instrument companies.

When Don Anthony came to the lab as Vice President for Research, he introduced me to a bit of discipline in these major instrument purchases. By demanding that we look at the benefits to the company of making each purchase (of course these were hypothetical, but we were forced to ask to what problem we might apply them and what a solution to this problem or problems would be worth) and then looking at a base case where we didn't buy the new instrument but tackled the problem with existing equipment, we could compute a return on the company investment, known in business as an IRR, investors or internal rate of return. Of course, the company did this for major capital projects, indeed every company does that to this day, but we were probably the first, maybe the only one to do it for purchasing of analytical equipment. It was a very good and useful discipline, it made us ask hard business questions even in a research and service lab setting, and I can't think of any examples where it led to the wrong decision.

Now I mentioned that when I was interviewed by the three lab directors they told me that the job of R and D was to make all the smaller businesses profitable. Indeed, beyond the businesses which had been acquired by Sohio, mainly as part of the acquisition of Kennecott, there was a whole activity of Sohio Ventures, supporting several very early stage start-ups that had originated with work at the lab. These included such things as growing milkweed for fibre to make disposable nappies, new batteries for power tools, a variety of other electronic ceramics businesses, etc. And while these were very interesting scientifically, and from the point of view of early-stage engineering to scale up and manufacture, as well as having some of the most interesting people working on them, for some odd reason I was not attracted to them. I was in a business setting, so I thought I should look to where the business action was. And I could see from my experiences that it was in the oil businesses, both exploration/production and refining/marketing. Now I didn't know much about the former, the so-called upstream business. I thought for that one really needed a background in geology. But refining was largely chemistry, and the throughputs were so great that a tiny improvement meant a great deal of money. I was attracted to materiality.

So it happened that when it was time for me to move out of my introductory role, I told Don Anthony that I wanted to work on these refining/marketing issues, and he arranged for me to become R and D director for Sohio Oil, the refining and marketing business. So great was the company's neglect of this business that no one had actually held this role. Not long thereafter I was seconded to BP in London, where I was to remain not for the two years originally promised, but until I retired some fifteen years later. A big chunk of my time in London would be spent on the technology opportunities for both the refineries and the oil products businesses. To some extent, my scientific life became my science and engineering life, though of course I still had my scientifically trained brain, even if I didn't make much use of my 10,000 hours of quantum mechanics after this point!

The Measurement of Octane: Using Advanced Computational Methods to Simplify Determination of the Most Important Property of Transport Fuel

ONE OF THE FUNDAMENTAL PROPERTIES OF GASOLINE is its octane (there is a comparable property for diesel called cetane). This property determines how to burn the fuel at maximum efficiency. As the fuel–air mixture is compressed in the cylinders of the engine, by motion of the piston, a spark is ignited at a certain point and explosive combustion occurs, forcing the piston down and driving the vehicle forward. If the temperature of an air–fuel mixture is raised high enough, the mixture will self-ignite without the need for a spark, and this is called the self-ignition temperature. Self-ignition is not desirable in an engine, because it leads to all sorts of pressure pulses in the chamber. The motorist recognises these as the sound known as knocking. Nowadays, with elaborate computer control of engines, and with high-quality fuel available everywhere, it is rare to hear the sound of knocking, but at one time it was very common. It was recognised early in the development of the internal combustion engine that different fuels had very different self-ignition characteristics, and that this depends on the chemical composition. An arbitrary scale was developed, the octane scale, in which the combustion of the pure liquid iso-octane

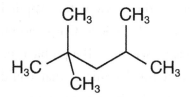

Iso-octane, octane number 100

was rated as 100, and that of normal (that is, straight chain, not branched) heptane

$$H_3C - CH_2 - CH_2 - CH_2 - CH_2 - CH_2 - CH_3$$

was rated as 0. At the time it was thought that no chemical or mixture could be higher octane than iso-octane, but that turned out not to be the case.

Now the measurement of octane was, in the mid-1980s, being done by having a certain standard engine that had been designed in the 1930s, whose compression ratio could be varied as it was injected with fuel mixtures that were produced in the refinery. In fact there were two measurements normally made, with the engine running under different conditions, yielding values known as the motor octane and the research octane. The octane number shown at the pump where you buy your fuel is the average of the two, in principle reflecting the performance at lower and higher speeds.

When I came to the Sohio research centre we began to think about new ways of measuring octane. Along with our other thrusts to automate measurement at our refineries and chemical plants, we wondered whether the entire engine method could be eliminated. Unknown to us at that time was that the same sort of work was already in progress at the BP research centre at the Lavera refinery and chemical plant in southern France.

Shortly before I came to Sohio, I heard a talk in New York, at a meeting of the local chapter of the Society for Applied Spectroscopy, by Marvin Margoshes of Technicon Corporation. Marvin was a friend, and a very clever scientist. The problem he tackled was to build an instrument for Technicon that could measure the protein content of grain. At that time there was a massive trade in grain (especially wheat) around the world, and the price was in part based on the protein content. Proteins contain nitrogen: they are made of amino acids that are nitrogen-containing compounds, so the standard approach had been to just analyse the total nitrogen content of a wheat sample and use that as a proxy for the protein content. Some unscrupulous grain dealers had taken to perverting the system by spraying their wheat with fats that were nitrogen-containing, so as to artificially inflate the apparent protein value, and hence the price.

I had worked for most of my career on the infrared spectrum of substances, and it is possible in the infrared region of the spectrum to distinguish proteins from fats, but the trouble is that the absorption of infrared radiation by these grains is very strong, so you would need to prepare a thin film sample, or a dilute solution, to make the measurement. And then there would be the question as to whether that very thin sample was representative of the whole. Sampling is always a problem with real-world measurements.

But Marvin Margoshes and his colleagues realised that there was a neglected region of the spectrum, the 'near infrared' (which is the high-frequency end of the infrared region), where the same information was present but the absorption was much weaker, indeed it was about 1,000 times weaker. He investigated various sorts of instrumental approaches in this region and found one where no sample preparation was required: simply dump some grains into a tube and make the measurement. Now it was not at all clear in this region of the spectrum which were the absorptions that could quantitatively give you the protein content, but here came another innovation. If you had accurate determinations of protein content of, say, 100 grain samples, and you measured their near infrared spectra, a computer could learn which bands were the ones whose intensity correlated with the measured protein content. Once you had a reliable correlation, you could move on to unknown samples. Of course, from time to time, there might be a new variety of wheat that would not be known to the computer, but then the learning just had to be extended. In this way Technicon had developed an instrument that could be placed in grain elevators around the world and stop the cheaters.

Well, that was interesting, and clever. It didn't fit with anything I was working on at the time, but of course it was filed away in my brain as part of my 'things that are worth remembering' file. When I went back and looked for an article or two about this, which I could put into my files of miscellaneous things I might want to come back to later, I found that Marvin's co-authors were Ed Stark and Karen Luchter. Ed I did not know, but Karen had been my classmate in graduate school at Purdue.

When I came to Sohio, my colleague Ken Gallaher and I started to think about octane measurement, and we turned to near infrared to see if it would work. At the same time our colleague Mike Markelov was trying to use near infrared for our chemical plant making acrylonitrile and other products. This made sense for two reasons – as I have said, if the absorption of the radiation is weaker, the path length for the beam through the sample must be greater. Now for online measurements you want a long path length so that it does not disrupt the flow of the sample in the factory. We also believed that, with the near infrared region of the spectrum being adjacent to the visible region, we could get optical fibres that could be put into the pipes in order to transmit the data back to the instrument, and this turned out to be true.

Besides the convenience of doing a simple measurement by flowing liquid through a glass tube for online measurement, compared to drawing a sample and taking it to an engine for a lengthy test, there was a good economic motivation for trying to replace the engine test. Octane is a key property of gasoline sold at service stations, and customers choose a grade of gasoline; in the past they also might have chosen the brand, based on the octane of the product. Regulators in each country regularly sample fuel from random pumps to check that the octane is at least at the level that is represented on the pump. To be sure of being in compliance, if a fuel is being sold as, say, 89 octane, most companies will ship something like 89.5 octane. This allows for any inaccuracy or imprecision in the testing, as well as any degradation leading to loss of octane between the refinery and the pump.

What do I mean by inaccuracy or imprecision, you might ask? Well, there is a true value to the octane, and a well-calibrated instrument, or engine, will give you that value. But if the instrument has not been properly calibrated, say by running well-established standards, a sample with actual octane of 89 might give a reading of

89.2. That is an inaccuracy. But even in a properly calibrated instrument, if we measure the same sample ten times, not every sample will give a reading of 89. We might find values between 88.9 and 89.1, for example. This variability is just due to how the engine performs, just as we are variable in how we do things. This is the precision of measurement, that is, we can say that the value of the octane is 89 ± 0.1. Now to deal with this sort of inaccuracy and imprecision, particularly the latter, we would always blend the fuel to a higher number, just to be on the safe side. But octane is a cost to the oil company, i.e. high-octane fuel costs more to make, and the difference between the specification on the pump and that of the actual fuel is called in the trade 'octane giveaway'. If the near infrared measurement could have better accuracy and better precision we could reduce octane giveaway, and hence make money. This turned out to be the case, but only once the method became the standard, and the regulators were using the same method. Of course, we still had the issue I discussed in the section on laboratory robotics, of inter-laboratory variation in measurement.

There was another interesting cost savings that emerged from our work. Different grades of gasoline are sent through a pipeline from refineries to various terminals, where additives may be introduced to them, then they are put on tanker trucks for delivery to filling stations. Now these different grades have different octane ratings, and in a pipeline there is an interface between the slug of one grade and of the next (one of these slugs could be a few miles long!). But because of the tight regulation of octane, you have to worry about mixing at the interface; some of the lower-octane product will blend with the higher octane. To compensate for this, the pipeline operator 'cuts out' a chunk of liquid from somewhere near the end of one parcel to somewhere into the beginning of the next.

They were doing this through experience, and cutting out enough so that it would be safe to specify the octane of the entire parcel. The material that was cut out was not wasted, but it had to be returned to the refinery for reprocessing. But now, if one could make an online measurement, very quickly, of the octane in the pipeline, the cut could be done with great precision, and the amount cut out greatly reduced. This sort of mundane-sounding improvement in an old industry can be worth a lot of money.

So we had a clear commercial motivation for our work, not just an inclination to do something clever. With this motivation we began to look at the problem of measuring octane. Sure enough, the spectra in the near infrared region of samples with different octanes were different, and we began to identify which absorptions were most sensitive, so we could try to use their intensity to build some equations that would predict octane. We thought, and indeed we found, that three or four specific frequencies would be sufficient to accomplish this.

We had made considerable progress before we found out that our colleagues at BP in France were trying the same thing. But they used a different approach. Being more mathematically inclined, they found a way to use all the frequencies in the spectrum to do the prediction of octane. There is a technique called factor analysis, which the French love. It is a somewhat peculiar thing in science that certain techniques, as well as approaches to scientific work, take on national characteristics. It is not totally surprising, because most researchers, after all, are trained in their home country, especially in the great science countries of the world. Faculty produce students, graduates, in their own image, and we can how see how this propagates through the science they do.

Now with factor analysis, you let the computer determine which of a large number of factors is most important, or, to put it more scientifically, correlates best with a particular property. I recall discussing this with my colleagues in BP France once, and they told me that the Government even uses it for the siting of nuclear power plants. Some communities strongly object to having such a plant, others don't seem to mind all that much. What they did to figure out which would accept a plant without protest was to devise a long survey, perhaps 100 questions, asking people about whether they liked watching football, how often they went out for lunch on Sundays, whether the church was an important part of their village life, how they felt about nuclear power plants, etc. Then, after they had the results, they could remove the question about nuclear power plants and predict very accurately how a village would react based on the answers to other questions. They knew which factors were important.

So this was the approach our French colleagues took to octane measurement. While in Ohio we were concentrating on several specific frequencies, and building predictive equations, the basis of which we could understand from our scientific knowledge as spectroscopists, the French threw all the data into the computer, churned it, and out came their own predictions. They had no idea why those equations worked; nor did they need to, because they knew with a quantitative degree of confidence that they did.

Eventually we all met in Lavera, near Marseilles, to compare our results and decide a way forwards. Of course, in addition to the teams from Ohio and France, we had a delegation from London, who, while they had not done any of the work, thought they should be involved. And then there were various executives who had paid for the work, on both sides of the Atlantic, who also thought they needed a trip to southern France. For two days we met and compared results. It was before the dawn of PowerPoint, and we each had piles of overhead transparencies with our data. Challenges went back and forth, and each of us was able to parry them. It became clear that we had two viable approaches to measuring octane online, very similar in how the data were acquired, very different in how the data were processed to produce a result. Later, as we were waiting to fly back, one of my colleagues said to me that it had been the most exciting two days of science in his time in the company.

Today near infrared is an established commercial technique for measuring octane, and there are even standardised test methods for doing it. Of course, it was not for Sohio or BP to develop this as a commercial product. Rather we filed some patents, and then sold the whole thing off to an instrument company to take it forwards.

Project Sunrise: The Complexity of Lubrication and Biodegradability

W E BEGAN TO THINK ABOUT ENVIRONMENTAL ISSUES and our products quite early, compared to others in our industry. In 1990, eighteen months after my arrival in London, I became head of something called the Products Division of BP, responsible for the technology behind our fuels, lubricants, bitumen and related oil products. This was a major role with a lot of business at stake. My technical staff were excellent, and at the most senior level had a deep understanding of the needs of the business, the technology behind the products, and the supply chain, because we did not invent everything ourselves. This was especially the case with the cocktail of additives that went into both the fuels and the lubricants.

Not long after I took over this role, Tony Roxburgh, who was head of marketing, spoke to me about an idea for a biodegradable automotive lubricant. Why would one want such a thing? Aside from corporate positioning through advertising, in a world where 'green' issues were becoming increasingly prominent, we calculated that much more oil entered the environment from leaks of lubricant out of cars than had been spilled in Alaska by the *Exxon Valdez*! Everyone saw these black spots on their driveways and the floors of their garages. It was for the auto makers to stop the leaks, but we might have the ability to make them disappear!

We knew that the main oil used in our lubricants, which came from refineries, or in the high-performance lubricant that Mobil produced, Mobil1, was not readily degraded by bacteria in the environment. But we speculated, and subsequently were able to show, that a different 'base oil', containing the chemical group known as an ester linkage, would be attacked and degraded readily. So the challenge was: could we formulate a lubricant with an ester base oil and appropriate additives that would still be a superior-performing product in cars, but biodegradable?

Now here is the challenge of a lubricant.

First and foremost, it must do its job of lubrication, that is, it must act to prevent wear of the metal parts that are turning at high speed. This is the application of the scientific discipline of tribology, the science of the interaction between moving surfaces.

It must be completely stable and retain its effectiveness over a long period of time (possibly for the lifetime of the vehicle nowadays), over a wide range of temperatures.

Besides preventing wear on the surfaces (erosion) it must also prevent any chemical degradation of the surfaces (corrosion) even as some acids build up in the fluid, which is common over time.

This is a very demanding set of specifications. To ensure that a lubricant meets quality standards (which you see when you buy it as designations such as SAE 10W30), the oil and automotive industries have jointly agreed certain tests, in engines, that lubricants must pass before they can be sold as having a recognised quality. These tests are expensive and time-consuming to carry out, so one needs to be quite sure of a formulation before submitting it for testing. Now to all of these specifications we had added one more, that when spilled on the ground it would be degraded by bacteria present in the environment.

This project was codenamed Sunrise, and it occupied a good deal of my attention during 1990–2.

Besides the base oil, a finished lubricant contains several additives, usually supplied by one of two or three major manufacturers. A part of my role was to maintain relationships with these manufacturers, so that we understood what technological advances they were making, and they understood our needs. While some of the additive companies were completely freestanding and independent (e.g. Lubrizol), others were associated with oil companies (e.g. Paramins, with Exxon, and Adibis, which was a BP company). Those in the latter group had to maintain a strict barrier between their own oil company and the additive company, and likewise we could not favour Adibis over their competitors.

Some of the lubricant additives help keep the base oil from breaking down, which is important chemistry in itself, especially when the base oil was not one that is commonly used, as in our case. But others are competing to be on the surface of the rotating parts, whether to prevent wear or inhibit corrosion. It is getting the balance exactly right that enables you to pass the key tests.

The oil codenamed Sunrise failed several tests the first time through. The launch of a new product is planned many months in advance, and these test failures meant that pressure was piled onto the team formulating the product. We met at least once a week, in a group chaired by Tony Roxburgh, to understand where we were and what measures were being taken to perfect the formulation. Tony was wise enough to see that there was danger that the stress people were under could lead to integrity being compromised. He asked me to bring together one or two other senior scientists in the company and, independent of him, to review the results of the tests as they were carried out. Only when our little independent review group was ready to sign off would he release the product to the market.

I have not said much so far about scientific integrity, but it is a big issue. The Sunrise example was one such case, and the process we put in place ensured that our company did not cut any corners, and that what we eventually said to the market (but six months later than we had promised) was absolutely backed up by data. But sadly this is not always the case, either in business or in academic research.

Faculty members, graduate students, entrepreneurs in start-up companies, large company technical staff – wherever technical results are being generated there is pressure to be successful, and there are individuals who will cheat. There is no point in looking for a polite word for it. When I was a graduate student at Purdue we had a case of an elaborate fraud being perpetrated by a graduate student, one who was regarded as a brilliant student, which led to a publication being retracted and the PhD degree being taken away. Later, as a faculty member, I had a student who produced results that were too perfect. Experiments just don't look like textbook drawings. As a venture capitalist in my post-BP life, I saw several cases of technology companies that were seeking funding which, on close examination, were clearly based on faked data. Usually, when one probed these, we were pushed away with the excuse that they could not tell us more because of the need for extreme secrecy. Of course we have had the Volkswagen approach to meeting difficult targets for emissions reduction without compromising fuel economy, using software to beat the measurement system. I am not saying that cheating is widespread or endemic, but it occurs sufficiently often that one needs to be alert to this possibility. The smart business leader of a technical programme will know when he is pushing staff so hard that the only way they can see to please the boss is by compromising their integrity.

Our biodegradable lubricant eventually came to market, first in Austria, then in Germany, the two markets where consumers were regarded at that time as being most sensitive to environmental issues. But by the time it did, the business leaders had lost their enthusiasm for it, and it was not promoted very well. Moreover, the advertising authorities in Austria would not let us say, as we wanted to, that this product was 'Better for the Environment'. All that they would agree to was that it was 'Less Bad for the Environment', which is not such a persuasive strapline! Well, we didn't make any money from this exercise, but we learned many fundamental things about our lubricants, and we also learned just how good the competitor product, Mobil1, actually was, compared to anything we were selling.

Clean Fuels: Applying Chemistry to the Problem of Automotive Exhausts, and Using the Science of Measurement to Optimise Data Collection; Chemistry Meets Politics

IN THE EARLY 1990S MUCH OF MY life was dominated by the science of what came to be known as clean fuels. When I became head of the Products Division of BP Oil, our fuels focus was completely on higher-performance fuels, how our fuels could make the driver feel more powerful behind the wheel. We were also concerned about engine cleanliness, in part because this improved performance, but also as a technique to counter the rise of fuel being sold by stations set up in the parking lots of the large supermarkets. We knew that these fuels, which were being sold more cheaply than ours, did not have all the additives present that cleaned the engines (effectively detergents in the fuel), and so we tried to formulate so as to make this a way to prevent our customers, who cared about their cars, from deserting us towards a lower-priced offering.

But quite suddenly the focus shifted, both in the US and Europe, to the environmental issues around fuels. Now environmental issues can mean many different things, but in this case it was mostly about local air quality – emissions of the oxides of sulfur (SO_x) and nitrogen (NO_x)[*] as well as particulate matter, both visible as smoke and the very small, invisible particulates that could potentially be very damaging to health. By this time cars were already fitted with catalytic converters that reduced the levels of carbon monoxide markedly, and also had some effect on SO_x and NO_x. Lead had already been eliminated from fuels across Europe and the US. Now regulators, and some legislators both in Washington and Brussels, were stirring the pot on whether the emissions could be reduced much further, making the case that this was essential for human health, particularly in cities.

One of the culprits was sulfur compounds present in the fuel. Crude oil has varying amounts of sulfur in it (high sulfur crudes are called sour, low sulfur are called sweet), and some of this sulfur makes it through the refining process into the final fuel. Indeed, we treated the sulfur compounds almost as if they were an additive, though we had not added them, because they helped contribute to a lubrication effect in the combustion chamber of some engines. The SO_x came from the sulfur levels in the fuel. By contrast the NO_x primarily came not from nitrogen compounds in the fuel but from nitrogen in the air that reacted with the oxygen at the high temperatures of combustion.

There was another problem. We knew already that sulfur compounds were a 'poison' for the catalysts in the catalytic converter. Catalysts are the materials, usually including precious metals like

---

[*] We use the subscript x to indicate that the emissions can be a mixture of SO_2, SO_3, SO_4, N_2O, NO, NO_2 and NO_3.

platinum and palladium, that help convert the carbon monoxide to carbon dioxide, and the NO_x back to nitrogen and oxygen. While it was true that the catalysts could recover, a steady dose of sulfur compounds in the fuel would reduce the effectiveness of the catalytic converter over time as well as instantaneously.

A third problem in cars related to what happened when a car started up from cold. The catalytic converters were really only effective when they were hot, and the heat was obtained from the heated exhaust gases passing over them. Now for a long journey this ineffectiveness of the catalytic converter might only affect the first 1 per cent of the journey. But most car journeys are short: trips to the grocery store, to pick up kids at school, to park at the train station. For these trips, which are the majority, the catalytic converter never got hot enough to perform its function.

We knew all this, and so did the auto manufacturers. But there was a mountain to climb that started with knowing some basic facts about emissions, agreeing these with each other, and eventually trying to reach consensus on how to meet ambitious environmental goals. In an ideal world we would find the optimum scientific and economic solution together, with government oversight, and implement it. Such a world does not exist. Instead we were in a confrontational situation, in which each industry would try to put as much of the technical and economic burden as possible on the other.

The problem was not just about private passenger cars. Of course, there are more of them than of any other kind of vehicle, but there were also important air quality issues from heavy-duty trucks, and their big diesel engines had very different interaction with fuels than cars. In cities there would be all sorts of delivery vehicles, taxis and urban buses. Each of these classes of vehicles had its own particular chemistry, and on the vehicle there were different systems for managing emissions. Moreover, it was not just new vehicles that we had to consider. If we were really to make some headway on emissions in less than the decade it takes for much of the vehicle fleet to change, progress had to be made in reducing the emissions of older vehicles, old trucks and cars, poorly maintained taxis, buses, etc.

In the US the auto and oil industries had started a joint technical programme, called US Auto-Oil, to address all these emission issues. One of our colleagues in Cleveland, Tom Bond, was very much involved in this, and on a trip to the US I consulted with him on how it worked. What I realised, though he did not say so explicitly, is that the auto and oil technical guys had become a bit of a club, maybe with the Government as their joint enemy, but they were certainly enjoying the work they did together, and their meetings at many places around the country. Less impressive to us in Europe was the relevance of the scientific work they were producing.

I was a bit on the periphery of this nascent European effort until I got a call one day from Rolf Stomberg, chief executive of BP Oil Europe. We knew each other, and I felt that Rolf respected me for my technical skills as well as my ability to navigate the BP system. He told me that he wanted me to lead for BP Oil Europe on a European Auto-Oil effort. By this time, my role in London had expanded, and I was Chief Technology Officer for BP Oil, so now covering refining as well as marketing. Our BP Oil chief executive, Russell Seal, was less enthusiastic about my taking a lead role in the European programme because of the time commitment, but Rolf was insistent. And I was more than a little interested. One of our German BP colleagues, Helmut Kuper, had been in the initial discussions, but Rolf felt, and correctly, that he was not senior enough in the company to be taken seriously by the other players.

I put through a call to a colleague in Shell, Bob Mackinven, and we talked, quite openly, about what had transpired thus far and where Shell wanted it to go, which was very much aligned with what I knew was BP's position at the time. It was pretty clear: the auto companies would find it much easier to comply with emissions regulations if we dramatically lowered the level of sulfur in the fuel. That was the main issue. But for us, this involved billions in capital costs to modify our refineries. By contrast, we felt that the auto companies could make modifications both to the software and hardware on the vehicles that would achieve lower emissions without any substantial change to the fuel, and we felt we had evidence that some of them were already achieving this, so why shouldn't the others be able to do likewise?

To deal with the problem of carbon monoxide on cold start, we faced a similar dilemma. There were ways that the auto companies could pre-heat the catalytic converter, without waiting for the hot exhaust gases to do so. This seemed to us like the best solution. But if you want to have less carbon monoxide, another way is to partially combust the fuel before it goes in the engine. Does this sound like nonsense? After all, the fuel is burned (combusted) in the engine to provide the power. The fuel is hydrocarbons, looking mostly like this:

$$CH_3 - (CH_2)_n - CH_3$$

And burning to produce carbon dioxide and water:

$$CH_3(CH_2)_n\text{-}CH_3 + O_2 \text{ (from air)} \rightarrow (n+2)CO_2 + H_2O + heat$$

It is the heat that expands the product gases, pushes the piston down and drives the vehicle forwards. If the combustion is not 'complete', instead of just CO_2 it also produces CO, carbon monoxide, and this is one of the things that the catalytic converter does, i.e. it converts CO to CO_2.

If you had some compounds that already contain oxygen, then these still can provide some power to the engine, but they will reduce the amount of carbon monoxide produced, because some of the oxygen is already in the fuel. The favoured choice for this was a compound called MTBE, short for methyl tertiary butyl ether:

In the US they were already starting to add this to the fuel. MTBE was made in chemical plants, relatively easy to do, but an expense for us in the oil industry. As it turned out, it had a brief but glori-

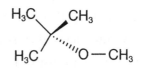

ous career as a fuel component, because it was found that some of it got through the engines unburned, into the atmosphere or onto the streets, and eventually wound up in the water supply. Unlike the unburned hydrocarbons, which didn't mix with water, MTBE dissolved in water. Before long, the concentration of MTBE in water was growing, people noticed the taste of it, and it did not break down quickly. This was one of several examples of adopting what seemed to be a simple solution to an environmental problem without thinking through the systemic consequences. It was also an example of why we were critical of the US Auto-Oil work.

There was another oxygen-containing alternative as a fuel component that could possibly meet the same objective, and that was ethanol. Ethanol for fuel was not a synthesised chemical or an oil product, but was produced by corn fermentation, hence a great business opportunity for farmers. To this day, support of adding ethanol, whose formula is CH_3CH_2OH, remains a crucial test for candidates for president in the US running in the early voting in

Iowa. I was once at a meeting as an oil company representative at something called the Governors Ethanol Coalition, the US state governors of corn-producing states who lobbied for standards requiring ethanol in fuel. By then our scientific work had established that the effects of ethanol on air quality were minimal, but that did not stop them lobbying for it. At that time, before several mergers and acquisitions, BP's fuel business was centred in the Midwest and the South in the US, and we were not hostile to ethanol, and willing to say so. The Governor of Iowa and I did a short TV spot together making this point, and then, as they often do with TV interviews, they switched off the microphones and asked the two of us to just chat so they could use this footage around the interview. So I said to him, 'You know, Governor, that this ethanol addition really doesn't do anything beneficial for the environment', and I smiled. He smiled back, as a politician a much better smile than mine, and said, 'You know, Doctor, I don't give a shit about the environment. I know what it does for the farmers of Iowa, and they're the ones who vote for me.' It was one of many lessons I learned during the clean fuels programme about politics and science not always being aligned.

The European Auto Oil programme came to be known as EPEFE, the European Programme on Engines Fuels and Emissions, and would come to occupy a lot of my scientific life in the early 1990s. David Price was Rolf Stomberg's government relations person in BP Oil Europe when it got going, and he had as an assistant a very hard-working and knowledgeable guy named Mike Wrigglesworth. We became regular telephone buddies, and co-conspirators. There was to be a scientific programme of work, which would be jointly undertaken and paid for by the auto and oil industries. On the one hand we were part of the oil industry group, but we also had to think about what outcomes would be in BP's best (or worst) interests. If, for example, the level of sulfur in diesel fuel was to be drastically reduced, some companies would have refineries that were better able to cope with that than others. Moreover, there were in Europe a number of big players in the oil industry – BP, Exxon, Shell, Total, Elf, plus some big global companies that were relatively small in Europe such as Texaco, Conoco and Mobil – and many smaller ones usually operating in only one or two European countries. An outcome requiring major capital investment might squeeze some more than others.

The auto industry association was called ACEA (something like the Association of European Auto Companies, rearranged so it either formed a pronounceable acronym or was based on the French word order) and the oil industry was organised around a group called Europia (European Petroleum Industry Association). We convened one evening in Paris as oil industry representatives, both to get acquainted and to plan our strategy, over a plush dinner sponsored by Elf (the French oil company that later merged with Total). I was going to come only for the meeting the next day, but David Price correctly informed me that the heavy hitters, as he put it, were all getting in early. The next day at La Défense in Paris we joined the 'Autos', and there were several hours of discussions of the idea of a joint programme and how it would be governed. Towards the end of this session, which involved another big lunch, there was a rather disorganised process for suggesting who would lead the various subgroups of the programme and who wanted to be members of these. I nudged Mike Wrigglesworth, who was sitting next to me, and said that when it came time to suggest a co-chair (one from Oil, one from Auto) for the research programme committee he put my name forward. Somehow, this process had not been agreed in

179

advance, and while colleagues from Exxon and Shell would surely have pushed one of their people, they were unprepared, and then it was too late. I was named co-chair of the research programme with the R and D director of Renault. I was now in the key position on the oil side to lead the experimental programme that would, hopefully, become the basis for advice to the European Commission and the European Parliament on what the future standards for fuels and vehicles should be.

Now in the science I had done during my career I was usually dealing with a specific problem of structure or reactions, and more recently I had been working on errors in scientific measurement, and on automation, because that was of great relevance to industry.

There is a large area of experimentation which is different in kind from what I had been engaged in up to this point, and while I was somewhat aware of this, it was pretty much as a chapter in graduate school texts that I had studied twenty-five years earlier rather than something I knew intimately. This is the area known as Design of Experiments. You may think that we always 'design' our experiments, and to some extent that is true. Is there a hypothesis, will the experiment test this, what is our anticipated result, what would be an unanticipated result? What we are talking about here as Design of Experiments is different. Supposing you have a very big space to explore and you want to know what's in that space. You cannot possibly make a measurement at every point; you need to sample within the space. The science of experimental design allows you to decide how many points to sample and tells you with what certainty you will know the contents of the space after you do the experiments. Clearly, if you want to reduce uncertainty you take more samples (which takes more time and costs more money), and again there is a science for how you increase sample density and how much you can learn by so doing.

This was what faced us in the European Clean Fuels programme. There were a range of fuel parameters, for example, for petrol (gasoline), the octane number, oxygen content, sulfur level, etc. Similarly, for engines there were factors such as compression ratio and engine size. And there were many different outputs, including the performance of the vehicle (this had to be defined in terms of a number or several numbers) and all the various environmental parameters associated with the vehicle exhaust – nitrogen oxides, sulfur oxides, particulate matter, carbon monoxide and greenhouse gas emissions.

It was not just how we did the science, i.e. using principles of experimental design to explore the 'space' of the multi-variable problem, it was the very nature of the problem that was different from what I had been brought up to do as a scientist. My scientific problems were well-structured questions that had answers. Indeed, we could assume that we knew the answer, do the experiment, and if it contradicted our assumption we would have to explain why. So it was with my bachelor's and doctoral thesis problems, and with the work I did subsequently on liquid crystals and on polymers.

Now there was not necessarily a well-structured problem, with a clear answer. We have many variables, we can change several of them, and perhaps there is an optimum result, that is, we might posit that there is a place in this space where we can achieve the lowest emissions to the atmosphere at the lowest cost. But perhaps there isn't. Maybe the lowest emissions have the highest cost, and it is not a scientific but a societal question as to whether this is a cost worth paying. More likely, and this is what was found, if we think of the results as a surface, mapping the emissions and the costs of achieving these, perhaps there is not a point where the optimum result lies, but it is rather flat, or there are several points, and one must decide what combination of the various parameters is preferred to achieve the result.

So yes, there were things we could change about the fuels, and yes, there were things that could be changed in the engines, or accessories added on to the vehicles such as catalytic converters, particle traps, etc., and science tells us the effects that these various changes have. But science does not provide a clear decision, or even a route to the decision.

Moreover, the auto industry and the oil industry each looked at this through their own eyes, and I think never looked at it through each other's eyes. Nonetheless, we did learn some things. It seemed pretty clear that reducing the sulfur level in both gasoline and diesel, at least to half what it was on average at that time, say to 100 parts per million (ppm), was a very good thing. But going lower than that, was costly and had only marginal benefits. It was interesting that especially in the test we did on big truck engines (so-called Heavy Duty Vehicles), the effect of varying the fuel quality was much smaller than we (in the oil industry) expected. To some extent this was also true in the latest car systems. Why could this be the case?

It turned out, and we now know this very well from the recent Volkswagen emissions tests, that the computers on the vehicles, the so-called engine management system (EMS), can adjust the engine operation, and peripherals such as the ratio of fuel to air going into the engine, to compensate for many things. So if you are accelerating, or braking, rather than running at a constant speed, the EMS senses this and changes various parameters in the engine operation to keep emissions low while maintaining vehicle performance. Likewise, if the EMS senses that the fuel has oxygenated compounds in it, or very low levels of sulfur, it adjusts the engine to give more performance, because it can do that without violating emissions constraints. That's why the surface of emissions seemed flatter than we had expected, because the computer was in effect flattening it. While we were puzzling over these results, someone from Honda remarked, 'Look, achieving low emissions is 90 per cent software and only 10 per cent hardware.' I noted that remark then, and it was the first thing I thought of when I heard about VW's problems in 2015.

One day, late in the process, as our results had all come together and were pretty much analysed, though the report was not yet issued, we were summoned to a large meeting at the European Parliament, where all those members of the Parliament who were actively involved in considering the new standards for vehicles and fuels were present. I was one of the key spokespeople for the oil industry, and presented the results for the sulfur content in gasoline and diesel. I thought I made a very coherent, sound scientific case for some lowering of the sulfur level, but nothing terribly dramatic or costly. There were questions, mostly polite, though it was clear that a number of the MEPs were inclined to a view that the oil industry needed to be punished for environmental damage. As I continued to defend us with the scientific results, frustration increased. Finally, Heidi Hautala, the Green Party MEP from Finland who was leading on the issue, interrupted me and said, 'Dr Bulkin, I think that you do not realise that the sulfur level in European fuel is not a scientific issue, it is a political issue.' Well, that was a big lesson for me, and she was, of course, correct. First, because science didn't really point to a 'right answer', and second, because scientists were not elected to decide such things. But it was a significant growing-up moment for me.

The second big lesson was that, while when we had concluded the study, the auto industry guys were depressed and the oil industry ones were fairly jubilant, because we felt that the science gave us the answer we wanted, the oil industry was completely naive in thinking that the process was finished. The auto industry realised that it had just begun. So while we more or less shut down our effort, they were just ramping up theirs, with lots of lobbying, press conferences, discussions on jobs and economic impact, etc. You can win the science and lose the prize, and we did.

The third lesson came after the regulations were set, and is an important generalisation across most environmental regulation. On both sides we had been asserting how very expensive it would be to comply with strict legislation, such as ultra-low sulfur levels in fuel. Where did we get these costs from? Well, it was not necessary to invent something new in order to meet these levels, there was existing technology and we were using the cost of that technology to project the overall cost to implement new standards across an industry. But once the levels are set, then everything changes. Instead of going for the textbook approach to compliance, the technologists and business people work together to find something better. This is especially true because there is no rush. We were setting standards that would come into effect eight to ten years later. If you set the bar high, and give industry time to comply, they will always come up with a better way of doing it. This principle, which was included in a famous article by Michael Porter and Claes van der Linde in *Harvard Business Review* in 1995,[*] surprised me then but still seems important to me now.

---

[*] Michael Porter and Claes Van der Linde, 'Green and Competitive: Ending the Stalemate', *Harvard Business Review*, September/October 1995, pp. 120–37.

25

Mn

Manganese

Clean Fuels 2: The Thirty Cities Programme, in Which Our Chemical Knowledge Allows Us to Change the Competitive Game, and Also to Explain the Impact to the Public in a Very Simple Way

WHEN I WAS RESPONSIBLE FOR REPRESENTING THE company in all the clean fuels work in Europe, and had overall responsibility, as head of technology for BP Oil worldwide, for our role in the US clean fuels work as well, I had very clear instructions from Russell Seal, our CEO of BP Oil, as to what my response should be to any of the frequent questions I got from the press about BP's policy: I was to say, 'We are in agreement with the policy of the oil industry on these issues, so if you want to learn more talk to our industry association.' I made this comment several times.

In 1995 John Browne became CEO of BP, and sometime after that he probably heard me make this response, perhaps in an investor presentation. A conversation ensued, in which John made it clear that he was of the view that we would decide what our position on issues such as clean fuels and climate change was, and that we were not to hide behind the industry associations. It might be, after deliberation and discussion, that we had the same view as the industry consensus, it might be that we had a different view. But we were, after all, grown-up enough and intelligent enough to formulate our own views.

The BP discussions and positioning on climate change are well known and have been documented elsewhere, and although I was a part of much of this work, it is, I think, more a part of my business life than of my scientific life. But clean fuels was very much a technical issue.

We wrestled with aspects of this for a while, but sometime in late 1997 John Browne and Rodney Chase, who was deputy CEO at that time, asked me to make a presentation to them, and other managing directors, about what we could do on clean fuels as a company if we wanted to do something to separate ourselves from the industry.

Naturally, I gave this a lot of thought and came up with a number of possibilities, some more ambitious than others. But I pointed them to the key issue of future legislation, which was going to mandate lower sulfur levels in the fuels, though probably not required for another seven years or so. If we could do something about this earlier than legislation required, much earlier, it would have a genuine impact on air quality, and set us apart from the industry. However, I also made it clear that this was not easy, indeed it was pretty much impossible to do across the complete geography we served, because so much of the base fuel we sold was obtained from the refineries of our competitors. Usually the way we differentiated ourselves, to the extent that we did, was by putting additives in the fuel at the distribution point; however, that would not deal with this problem. But I felt that there was still one possibility. There were, across our network, a number of cities where we controlled the quality of the fuel, because we were the sole source of supply from one of our refineries. We could produce some low-sulfur fuel at these refineries, just for our own use in the target cities, and we did not have to give this low-sulfur fuel to our competitors. There was a spirited debate, but in the end it was agreed.

I came up with this with only minimal consultation with the people on the ground who would have to do the work, a mistake in some ways, but as I was presenting several alternatives it was probably best not to panic anyone into starting off in one direction only to hear that management had chosen another. But now we had to crystallise the germ of an idea into a programme to be rolled out.

During 1998 I worked on finding cities that our refineries could supply with a differentiated, low-sulfur fuel, either gasoline, diesel or both. It was not easy, because the sulfur had to go someplace. We might be able to reduce it by increasing the sulfur level in some other product, for example the fuels we supplied to ships, but that was not really a good option (though better for local air quality, as pollutants into the air at sea were rapidly dispersed and had little impact on human health), or we could take it out as elemental sulfur and hopefully sell that as a product to the chemical industry. For each city we explored, we looked at both how to do it and what it would cost to do it, and then I devised a simple metric to aid in decision making: If the cost was $x, how much additional market share would we have to get in order to pay back this cost? Generally, in the markets in which we operated, our share of retail fuel sales varied from about 4 per cent to as high as 25 per cent. The low market share places were particularly attractive, because they tended to be highly fragmented, and if, for example, one could move from 4 to 6 per cent that was a big gain. Moreover, research had shown that while it was very hard to gain new customers, once you did it was just as hard for others to take them away.

I had a list of cities where we could segregate our supply of fuel, and, working with refineries in the US, France, Germany, and Turkey, we knew we could move ahead in several places. In principle we should have planned out a programme in a very systematic way, prioritising cities, etc. But sometimes events take control.

John Browne agreed to give a talk at the Economic Club of Detroit in January 1999. What we were doing was very much connected to our relationships with the auto industry, and would be viewed very positively by them, for, as I described earlier, putting a very low-sulfur fuel in the market made their job easier. And now we were breaking with the oil industry, which was still screaming, after all these years, about how hard and how expensive it would be to produce such a fuel. So what better place to announce this than Detroit? One of my jobs in BP was to manage our relationships with the auto industry, and I had developed a close working relationship with very senior people at General Motors and Ford, as well as many of the European manufacturers, especially Mercedes and Volkswagen.

Nick Butler was the main speechwriter for John, and we began to work on the Detroit speech. At this point I was still vague about how many cities we could do, when we could do them, and in what sequence. The speech forced me to focus on this. At one point Nick pushed me hard as to how many, and in the course of our back-and-forth discussion the number thirty emerged: the Thirty Clean Cities programme was born.

In due course John Browne and I flew to Detroit on a cold January day, and we met with Harry Pearce of GM (who was receiving treatment for cancer that suppressed his immune system, so our meeting was by phone), Bill Ford of Ford and the CEO of Chrysler. Then John spoke at the Economic Club of Detroit, and we announced the Thirty Cities programme, namely that we would be introducing, many years ahead of legislative requirement, ultra-low-sulfur gasoline or diesel in thirty cities worldwide. We did not say which cities, although we knew by this point which the first ones would be, because we still wanted to keep a competitive edge.

I had been working with our Washington office in preparation for the speech, and I knew that they had been working on a surprise. Throughout the talk I could see Mike McAdams from the Washington office pacing around in the back of the room, occasionally running out and coming back. The speech was concluded and John was answering questions. Traditionally at the Economic Club of Detroit, one of the dozen or so local high-school students who are invited to the talk gets to ask the last question, and I remember it very well. John had recently been knighted, so he was Sir John Browne, and the student asked, 'What did you have to do to become a knight?' to which John answered, 'I don't know and I hope I never find out.' At this point the chairman rose to thank John for the speech, and I saw Mike McAdams coming forward to the lectern from the back of the room, and he handed the chairman a piece of paper. He read out a message from President Clinton congratulating BP on its clean fuels initiative. The next day we were front page news in *The New York Times*. For the first time in my scientific career, the work I was doing was changing the game in a major industry.

The second event was of a more scientific nature. We agreed that our first three cities would be Paris, Istanbul and Atlanta. As we were preparing for Atlanta, I was feeding the press office specifics of how much various pollutants would be reduced in the air in Atlanta because of our new fuels. This required some approximations, but I had got good at doing this sort of calculation. The problem was that the press people, correctly, said that people could not relate to tons of nitrogen oxides, things like that. So I thought about it, and I think I was the first to invent a kind of horrible metric that is now used quite universally. Instead of talking about the tons of pollutants that would be avoided, I calculated that our actions would be the equivalent of taking 24,000 cars off the road. Now that was something everyone could understand. How did I get to such a number? Of course, there is no formula. I started with the amount of NO_x emitted by a single car and used my old friend, dimensional analysis, to work my way forwards to the number of cars equivalent to the reduction in emissions.

The third event that affected the entire thing almost cost me my job, and while that would not have been a tragedy, there was still one more good job ahead of me in BP, though I didn't know it at the time, and it would have been a shame to miss that.

At around the same time as we were working on our clean fuels initiatives, BP announced its merger with Amoco. The oil industry was beginning what would become its biggest reorganisation in decades. A few years earlier there had been a merger of the refining and marketing businesses of BP and Mobil in Europe. This was a big deal at the time, but it was nothing compared to what came next. After BP and Amoco merged, so did Exxon and Mobil, Chevron and Texaco, Conoco and Phillips, Elf and Total.

When the senior posts were allocated in the new BP Amoco, the refining and marketing business was initially co-led by Doug Ford of Amoco and Peter Backhouse of BP. Peter was a good friend and supporter of mine, and very close to John Browne. So in principle I had full support for my clean fuels programme. But Doug Ford was not on board at all. We had a meeting of the leadership of the business, I presented a rational business case; he was not convinced. As we came closer to the time of the Detroit speech, Doug tried hard to weaken the commitments that John Browne wanted to make, and I found myself in the middle. This should not happen in industry – the boss was clearly John Browne, and he knew what he wanted. But this was immediately after a giant merger and everyone was protecting everyone's feelings.

Doug could not do anything about the programme, but he could do something about me, or he thought he could. He called me in with all sorts of accusations about things I should have done but hadn't, in terms of keeping him and others informed. I was sure at the end of one of these conversations he was going to fire me, but he either lacked the courage or the authority to do so. Things rumbled on and we did the Detroit speech, and then prepared for our first three city launches, in Istanbul, Paris and Atlanta. Rodney Chase, at this time Deputy CEO, and also a supporter of mine, did the Paris launch along with me and Michel de Fabiani, president of BP France, and was going to do the Atlanta one with me.

We flew to New York first for some meetings, and then down to Atlanta. There were several of us heading out to the airport, so we had two cars. Rodney arranged that he and I would ride together. In the car he told me that Doug Ford wanted to get rid of me, but that he and John were not going to let that happen. In addition to my work on clean fuels, I was also heavily involved in the climate change work at BP at the time, again very strongly supported by Rodney but not by the Amoco heritage management. So I was saved for my next role.

Clean Fuels 3: Beyond Petroleum, Reviving Some Very Old Ideas and Inventions Becomes Possible

W HEN A DISCIPLINE REACHES THE MATURITY OF chemistry in the late 1990s, and attention focuses more on applications or extensions (into biology and materials) than on the basic science of the core discipline, one might think that the first place to look for these applications/extensions is around the most recent basic discoveries. After all, there had been decades, centuries even, to mine the depths of the oldest known reactions.

This turned out not to be the case, in large part, I think, because the game changes, and what was important in a winning strategy in 1910 or even 1960 is very different from what is important in 2010. For example, environmental issues push customers to demand different things in their vehicles. Or technology changes – for example, more sophisticated electronics make solutions feasible that were previously very difficult to implement. So while we were making the changes to reduce the sulfur levels of gasoline, and do other things to produce cleaner burning petroleum fuels, a new thought was emerging, particularly in the car industry, much of it rooted in very old chemistry.

When mass production of automobiles was beginning, in the early part of the twentieth century, there was actually a series of battles over what would be the best engine and the best fuel. Internal combustion (because the fuel burns inside the cylinders in the engine) spark ignition and compression ignition engines, that is, those that we have today running on gasoline and diesel fuel respectively, were only two of the alternatives. Ethanol made from agricultural products was considered as a fuel for those engines even then, and almost won out. Remember that 100+ years ago the oil industry was still very much in its infancy, and no one knew how much oil there might be. Cars using steam engines – that is, external combustion engines where the fuel is burned outside the engine, used to make steam, and the steam provides the motive power to the car – were manufactured. If you go to a rally of very old cars even today you can sometimes see a Stanley Steamer or other more than 100-year-old steam engine cars. Since batteries already existed at that time, electric cars were tried. Indeed, electric cars battled for dominance with internal combustion engine cars, the oldest commercial ones being made in the 1890s using batteries and electric motors, and were used in many urban taxis. As was still the case in the late twentieth century, the issue then was one of battery range and weight.

Way back in 1828, William Grange invented something called the fuel cell. Now I have previously referred to the thermodynamics of making hydrogen and oxygen from water, and how this required putting in energy in the form of electricity or heat. Most people who took a high-school chemistry course will have seen the experiment of the electrolysis of water, in which an electric current is passed through water leading to the production of hydrogen (H_2) and oxygen (O_2) gas. If a spark is passed through a mixture of the two gases they combine spontaneously to form water, with a great release of energy. What Grange saw was that it was possible to get the energy out from the combination of hydrogen and oxygen in a controlled way, as electricity, so if one reaction is water + electricity gives hydrogen and oxygen, then it is possible to do the reverse reaction, hydrogen + oxygen gives water plus electricity. The device that does this is called a fuel cell. It hung around in the literature of chemistry for about 150 years until NASA scientists realised that this would be a good way to generate electricity in space. In the late 1990s fuel cells began to be re-examined by the leading automotive companies as an alternative to gasoline- or diesel-fuelled internal combustion engines. Surprisingly, this came before the renewed interest in battery electric vehicles, although the thrust was the same: produce clean, efficient vehicles without use of oil products, though there is a big question about where to get the hydrogen from in the case of fuel cells. There were also unknowns about how much batteries could be improved, and on this the economic challenge that fuel cells faced from batteries.

With my accountability for relationships with the auto industry I had developed a relationship at a very senior level with General Motors, and while they had a nascent programme on fuel cells based in Germany, the real action was at Daimler Benz. I was also able to have discussions with them at a very senior level because Rolf Stomberg, who was my boss at the time, had the foresight to develop a friendship with Dieter Zetsche, who would later run the entire company. My colleagues and I travelled to their headquarters in Stuttgart several times for discussions. We repeatedly came back to the problems of making hydrogen. One idea being pursued was to make the hydrogen on board the vehicle, by having a small chemical reactor that converted methanol, a readily available, cheap liquid, into hydrogen and carbon dioxide. The reaction is:

$$CH_3OH + H_2O \rightarrow CO_2 + 3H_2$$

Again this was old chemistry, but several groups pursued it, because it was attractive to be able to continue to fuel a car with a liquid fuel (methanol) rather than gaseous hydrogen. Like many big ideas, on which lots of money is spent, this turned out to be a dead end, because of some basic chemistry problems, of which the most serious was that while in principle the reaction shown above went completely to hydrogen and carbon dioxide, there was usually a small amount of carbon monoxide left in the mixture. As chemists (rather than automotive engineers) we knew this was bound to be the case, because we know how hard it is to drive any chemical reaction to 100 per cent completion. Now if this carbon monoxide, CO, got into the fuel cell, it attacked the platinum catalyst that made the conversion of hydrogen and oxygen to water plus electricity happen under very controlled conditions. And once that catalyst was 'poisoned' the effectiveness of the fuel cell dropped. To remove every bit of CO would be too costly.

Gradually attention shifted back to storing pure hydrogen on the vehicle. To achieve a range of 300 miles, say, requires that the hydrogen be stored at high pressure. Or perhaps, some thought, there were solid materials that would absorb (either physically or through a chemical reaction) a lot of hydrogen, and then release it, by heat or some other trigger.

I felt that it was important that BP as a corporation understand whether this was a serious threat to our business, and on what time scale. I had to understand the technology, separate what was demonstrated from what was optimism, what would be a breakthrough and what would be a killer flaw, and I also had to understand the economics to see whether anything of this sort would ever occur. I had to do all of this even where some of the senior executives, particularly Doug Ford, who was running the business that would be most directly affected, evinced no interest in being educated about it at all. From time to time I gave little lunchtime talks in BP headquarters in an area where staff could gather. After one of these on fuel cells and hydrogen, attended by about 100 junior staff and personal assistants mainly, one of the few executives present said 'now the PAs of the managing directors know more about fuel cells than any of their bosses'.

Now, fifteen years later, we still have not had any mass production of fuel-cell-powered vehicles, though interest in vehicle electrification has grown dramatically. Toyota and others continue to invest in the technology, but there remain huge cost barriers to putting in place a hydrogen fuelling infrastructure. The oil industry relies, to some extent, on the barrier to innovation taking away its business being the laid-in infrastructure of refineries, distribution terminals and filling stations. Simple battery-powered vehicles get around this, because they use the existing infrastructure of electricity provision, but hydrogen does not.

Chief Scientist: How University Research at the Most Prestigious Universities, in Chemistry and Related Sciences, Shifted to Solve Major Problems, and This Is a Manifestation of the State of the Discipline

THE THIRTY CITIES PROGRAMME WAS ROLLING OUT around the world, and a team was dealing with all the climate change actions we had under way. I realised that I had less and less to do, and though I was still called on for advice on lots of things, in fact I didn't have a job any more, though I was still employed. Andrew Mackenzie changed all that.

Andrew was at that time an up-and-coming leader in BP, and today he is chief executive of BHP Billiton, one of the world's leading mining corporations. He had a strong technical background in geology, but had worked in a variety of roles across the company.

After a succession of older chief technology officers in the company, Julian Darley and David Jenkins, John Browne had appointed Andrew to this role. He put in place a structure of technology vice presidents for the various businesses, which still endures.

At some point Andrew approached me with an idea of a role for me. BP had already started to create a research activity at the University of Cambridge, called the BP Institute, with a focus on oil reservoir management. This mainly deals with how one recovers the maximum amount of crude oil from under the ground or under the seabed. In the early days of the oil industry, and for many decades, it was thought that no more than a third of the oil could be produced economically. Science had advanced, and these numbers were now pushed up to half, and could go beyond. Indeed, for a big oil company the game was said to have become one of 'recovery not discovery' in the late twentieth century. This was an exaggeration, there was still more oil to be discovered, something we held as a matter of belief but were to find out dramatically as fact in the early twenty-first century. Still, the activity at Cambridge, John Browne's alma mater, was just getting going, and a director had been hired, a very good young scientist named Andy Woods. Cambridge also was home to many outstanding earth scientists, including Professors Dan Mackenzie and Ekhard Salje. It was a logical place for BP to invest, both on technical and political grounds, given the importance of enhancing the company's reputation in the UK as well as the excellent work that would come out. Of course, it created sore feelings, at arch rivals Oxford and Imperial College, especially the latter, who felt that they had a leading position in petroleum geology.

Andrew had the idea that we might create several substantial research centres at universities, and that the next place to do something would be the US. Would I take this on? He had approached John Browne and been given assurance that up to $15 million could be made available for each of two centres. Possibly more.

Now, like other big companies, BP had a long history of giving money to universities, but it was mainly in small amounts, to sponsor a graduate student, or participate in some sort of multi-company sponsored programme around a topic of interest. In this way many of our technical staff directed some money back to their own universities, sometimes to the professor with whom they had done their thesis research. A bit of a welfare programme for academia. Like many other facile statements, this is too harsh. One such grant, to Professor John Meurig Thomas of Cambridge University, had led to a substantial business for BP Chemicals, probably on its own repaying ten years' spend of the entire university programme. And I am sure there were many smaller paybacks as well. When I rounded up information on all of this (surprisingly we did not have it in one place), there were more than 400 individual grants to universities, none of them very sizeable, and most seeming to have little impact or benefit for the corporation. I suspected that several had just continued without anyone taking a decision that it was something we really wanted to do. It was part of my mandate to clean this up, without upsetting everyone on both sides of the equation if possible.

Here was the beginning of a new chapter in my scientific life. For years as an academic I had been scrambling to secure funds, from government or industry, for my research. Now I could be the one giving out money, and rather substantial sums at that. But I had to make some judgements as to who should get the money, and these needed to be scientifically sound for sure, but it seemed important to me that the work that would be carried out be of deep interest to at least some of the technical community within the company, and ideally some of the business community as well. What I was about to do mirrored a lot of what was happening in chemistry as a discipline at this time.

There was a long time, most of the twentieth century, when every industrial lab with a reputation had a substantial basic research programme alongside its business-focused research and development. In BP this was called Corporate Research, and it had continued until 1992. In other companies it was called Central R and D, or something of the sort. Appended to these basic research programmes were university expenditures, all in support of basic research, of which a large chunk was in the chemical sciences. Some companies even created their own awards for the basic research areas of interest to them. In this way the whole chemical enterprise moved forwards towards the solution of fundamental problems, some of which, sooner or later, had applications in the chemical industry and related industries such as agriculture, food and pharmaceuticals.

This was especially true in the understanding of catalysis, i.e. how certain materials, usually solids, acted to speed up reactions. Most chemical processes used in both chemical plants and oil refineries use some sort of catalyst, and for decades these were poorly understood, developed largely through accidental discovery followed by empirical improvement processes. As new tools developed for looking at solid surfaces, under reaction conditions, we began to learn just how the catalysts worked, and used this knowledge to design new ones. This was a major thrust of university–industry collaboration during the last three decades of the twentieth century, continuing in the twenty-first century. It started to fill in with molecular-level understanding a big section of chemistry that had been poorly understood. That understanding remains incomplete;

it is one area of chemistry where we may still see major advances in our knowledge.

As I took on this challenge of major university programmes for BP, I thought it was time to realise that things had moved on in the enterprise, and we needed to stop paying for work on problems that were 90+ per cent solved, and tackle other things that could be of real value to the company. If universities were to be a part of this activity, then they would want to turn their focus to what we were interested in, or so I believed. This interest would reflect the emerging maturity of chemistry as a discipline. There remains a similar challenge for the Government entities funding basic research in chemistry.

While we had decided to give Cambridge money, and even decided the amount, and *then* worked with them to fashion something that we wanted to give the money for, I now took a different approach. I had a lot of discussions with colleagues. To achieve what I wanted in terms of doing something of interest to both the technical community and the business community in the company required having a strong network in both of those communities, and people who were always ready to talk frankly with me. So I had lots of conversations with colleagues before I started to say anything to the universities. By far my most important conversations were with Chris Gibson-Smith, at that time a managing director of BP, responsible for climate change initiatives among other things, and someone I greatly respected for his insights. At one point Chris said to me, 'Do something related to methane; that is going to be the most important molecule for us.' That idea stuck with me.

In the end, I settled on two main themes. The first seemed obvious to me – do something connected to climate change. Now at that time there was the beginning of interest in what we now call carbon capture and storage (CCS), and the first meetings were being held to discuss this. The idea is, in principle, straightforward. Some significant portion of the carbon dioxide that is going into the atmosphere and causing global climate change is concentrated in relatively few (that is, a few thousand) places, namely big fossil-fuelled electricity-generating plants. They burn fuel, for example coal or gas, and most of the carbon in that fuel winds up as carbon dioxide, through the same sort of processes as those I described earlier for the combustion of gasoline or diesel fuel. Now since the carbon dioxide in these power-generation plants is coming out of big furnaces, we have the possibility of separating it from other gases present in the smokestack, our old friends SO_x and NO_x, carbon monoxide and particulate matter, and then putting the carbon dioxide into a pipeline, perhaps sending it deep underground someplace, or even to the bottom of the ocean. This CCS process would require a lot of science and engineering if it was ever to be practical, and I wanted to use one of the university centres to advance this science and engineering.

I also felt, and this is important to creating such a centre, that much of the science and engineering could be useful even if CCS never happened. For example, there was new technology needed for separating the mixture of gases coming out of the furnace, and we have lots of applications for gas-separation technology. We also needed to compress the gas so it could be efficiently piped underground, and gas compression was still being done as it had been decades earlier. Perhaps there was an engineering breakthrough here which had lower energy requirements. And if we were going to explore the possibility of deep ocean storage of carbon dioxide as a liquid, there was a lot to learn about how the stored CO_2 would behave, and what the effects would be on the ecosystem at the bottom of the ocean. I wanted all this work to be set in a place where there was some deep thinking about climate change itself, so that the university scientists and engineers could influence our business leadership with their thoughts.

The second area was much more mainstream chemistry, and it is the area known as gas conversion. This was my response to Chris Gibson-Smith's suggestion. There is a huge amount of natural gas in the world, and it is very widely distributed. True, some countries – the US, the UK and Norway (the North Sea), Russia, Canada, Qatar among others – have a disproportionate share, but there is a lot of it. From where we stand today there is so much that we can say that gas seems almost infinite as an energy source for Earth. Now to get the gas, most of which is the simple molecule methane, chemical formula CH_4, from where it is found to where it is used, there are three choices.

Pipeline, and we have many such pipelines, for example from the North Sea to the UK and Northern Europe, from Russia to Eastern Europe and into Germany, from Texas up to the eastern US, that supply us with energy for cooking, heating and power generation.

The gas in the Middle East is often brought to market by the second alternative, *liquefaction*. The gas is chilled and compressed until it is a liquid, then loaded aboard tankers, and shipped around the world, where it is converted back to gas and put into local pipelines. This is a less desirable alternative, because it takes a lot of energy to liquefy the gas, additional energy to keep it cold for a prolonged period, and some more to ship it, but it is still economical.

The third option is to carry out a chemical process to convert the gas to some other chemicals, for example something that resembles gasoline or diesel fuel, and then ship the liquid fuel to market. This is of interest for so-called stranded gas, where you cannot easily send it by pipeline or liquefy it. An example would be the vast reserves of gas on the north slope of Alaska. With the decline in oil production in Alaska, there is space in the Trans-Alaska Pipeline for liquids that could be synthesised on the North Slope and separated out later.

Now this third option, chemical conversion of gas to liquids (GTL), originated back in the 1920s, not for natural gas but to convert coal to liquid fuels. It had been known for centuries (though the chemistry was elucidated only in the 1800s) that coal could be converted to a mixture of carbon monoxide and hydrogen using steam. The gas that was formed was known variously as water gas, town gas or synthesis gas. Some of the early discoverers of this phenomenon in the 1600s used the gas to light their homes. The name synthesis gas, known to everyone in the industry as syngas, came from the development of chemistry in the early 1900s showing that many different complex molecules could be synthesised from this mixture of CO and H_2. Indeed, when I studied high-school chemistry in 1955 I was required to memorise about twenty of these reactions, without the slightest hint as to why they might be important. I completely forgot about them until I joined Sohio in 1985, and found that there was a modest research programme there, and at BP in the UK, on this subject. But it is simple chemistry and easy to relearn. Besides, I now saw that it was the basis of some existing big industries, and potentially the basis of some huge new industries.

In the 1920s, two German scientists, Franz Fischer and Hans Tropsch, developed chemistry for converting syngas to long-chain hydrocarbons, essentially waxes that with subsequent processing using additional hydrogen gas yielded a mixture of chemicals very suitable for diesel fuel.

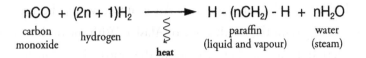

$$nCO + (2n + 1)H_2 \longrightarrow H - (nCH_2) - H + nH_2O$$

carbon monoxide   hydrogen   **heat**   paraffin (liquid and vapour)   water (steam)

This was to prove very important to Germany during the Second World War, as Germany had large reserves of coal but no significant amount of oil. By commercialising the so-called Fischer Tropsch process during the 1930s, Germany was able to have fuel for its army after it was cut off from supplies of oil. South Africa likewise produced liquid fuels by these reactions during the period of boycotts.

While in the 1980s there was still interest in coal gasification, the need to get gas to market intensified the interest in converting methane to syngas, and from there to diesel or gasoline. Other products were also of interest, for example methanol, which was already made via a commercial process from natural gas, and was a potential replacement automotive fuel (it is used in Indy racing

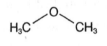

cars), and a compound called DME, dimethyl ether, which could replace diesel fuel or propane used in cooking.

There had been a lot of effort on GTL, particularly at our competitors, Shell, Mobil and Exxon. By contrast the BP effort had been modest. Nonetheless, despite all this effort there was a view that much more could be achieved in finding new routes to reducing the cost of making liquid fuels from natural gas. It was that challenge which I wanted to set for some university research groups.

My approach to both these problems, climate change with an emphasis on CCS and gas conversion, was to pre-select, from among US universities, a few places where I thought the most promising work was being done, and then have them compete for the funds by making proposals to us. It had been pre-determined that this was to be a direction of funding into the US by BP, so I did not look elsewhere, and inevitably this led to some hard feelings. I was also very determinedly elitist in my selection process. It is very clear in the US which are the first-rank universities, say the top twenty, the second tier, say numbers twenty-one to fifty, and the rest, when it comes to research. While there might be a small cluster of people doing excellent work in a university outside the top twenty, I felt that we would get the best work done, and the greatest benefit to the company, from funding research at a top-tier university only. Again, there were some bruised egos.

For the climate change programme, I was attracted to Stanford and MIT. At Stanford there was an outstanding geology effort, very mathematical, under the leadership of Lin Orr. They understood the oil industry, and probably were as good as any in the world at the physics of what happens below the surface. At MIT, there was a lot of work in progress on CCS already, with Howard Herzog one of the world leaders, and a number of other key scientists. Moreover, MIT was a strong centre of research on climate change, and BP had been supporting work there on climate change for years.

I was puzzled for a while about where else to look, when one of my colleagues suggested we consider Princeton. He had heard a presentation by Professor Robert Socolow which he found impressive. Socolow was a physicist with interest in environmental problems. While I hadn't heard of this work before, some literature searching confirmed that there were actually a number of people doing interesting work there. And I knew that in the chemistry department at Princeton my friend and long-time colleague Thomas Spiro had shifted his attention to environmental problems as well. So I settled on Stanford, MIT and Princeton as our three possibilities.

For the gas-to-liquids centre, the problem was really one of narrowing it down to three, and in the end I did not succeed at that. University of California at Berkeley was obvious, because work was underway under Professor Alex Bell, and his colleague Enrique Inglesia. Inglesia had come from Exxon to Berkeley, and we knew both of these guys well, and had even provided some smaller funding for them. Berkeley was certainly one of the top five chemistry research programmes in the US. Interestingly, Berkeley had a joint department of chemistry and chemical engineering, and the engineering flavour was important. Our second clear candidate was California Institute of Technology. Again, BP had been supporting work there, and we had huge respect for the work of John Bercaw, Bob Grubbs (a future Nobel Prize winner) and Mark Davis, younger faculty Frances Arnold (now also a Nobel Prize winner) and Jonas Peters, as well as such towering figures as Harry Gray and Rudy Marcus, both of whom I have already mentioned in other contexts. There was great depth of excellence in that department, which was also a joint chemistry-chemical engineering department.

After that my colleagues in BP had two strong recommendations, Illinois and Texas A and M. Both were top-tier chemistry programmes; Illinois was the third chemistry-chemical engineering joint department (and those three were really the only ones of note in the US with this configuration), while Texas A and M had been steadily improving over a few decades, and there were some key individuals there whose work impressed us. So, after a bit of agonising, we decided to include all four in the pool. There was one factor that tipped the balance for me on having Texas A and M in. Chuck Bowman, who had been a senior executive in BP, and always very supportive of things I was doing, had recently retired and joined the university in a senior capacity. Even if they were not going to win eventually, it was a courtesy to him to include them. These are the kinds of not strictly scientific judgements I had learned to make.

I developed what I thought was a very clear process. First, I called a lead person, out of the blue, at each university, and said we were thinking of giving away $15–20 million for research in a particular area, and were they interested in competing for this. I made clear that this would be a limited competition, pre-selected, so the odds were pretty good. When I called Rob Socolow at Princeton he was away, and I asked that a message be passed to him. Some hours later I got a call back from Texas, where he was travelling with his wife. He later told me that he had been inclined to ignore my call (since I had not said very much in the message) but his wife told him, 'Don't be stupid, you never know what this could be about.' She was right.

After an initial conversation about what we were thinking, probably an hour by phone, I arranged to go and visit each university, just on my own, to meet some of the faculty and talk to whatever group they wanted to put together about our plans, interests and what they needed to do. For this, I developed something like a script – not really a script, but the key points I wanted to make to each – so that, to be scrupulously fair, everyone got the same briefing. Of course, I wanted to hear from them as well, with their first thoughts, an idea of who would lead and who would be involved, hear any negatives or even hints of concern from the university administration about taking money from an oil company, and on the more positive side some idea of what the university was willing to offer in support of the effort.

Now it is interesting setting yourself up as a granting agency, with quite a lot at stake. In fact I had a little bit of experience with this. Some decades earlier, Carol Shaw of the University of Dayton and I were doing programmes for women trying to re-enter the scientific or engineering workforce, and we convinced a Federal Government programme to let us make grants to fifty different colleges and universities around the US, then train them on how to run the programmes. I recalled this experience very well, and thought at the time that this is like something you learn to do as a teenager in a summer job, and then get to practise in a real job later. Those grants that Carol and I gave were for $5,000 each, and now I was planning on $15–20 million. No problem, just a matter of scale.

So I now gathered two teams, just three of us on each, to go back to each university after an interval of about a month and hear their response to our invitation – what would they like to focus on within the broad topic areas we laid out. This gave a chance for the faculty to meet and debate, form some plans, know who they want to include, lean on the university administration to make some commitments in support of our grant – basically all the mainstream stuff that faculty are good at doing. I suspected that there would be some interesting infighting along the way as well.

The gas-to-liquids centre was eventually established as a joint effort of BP, Caltech and Berkeley. All four universities had interesting science, but Caltech and Berkeley complemented each other in their approaches, one trying to make the mainstream way of doing this chemistry much more attractive, consistent with the state of evolution of chemical processes at that time, the other trying something completely different that was higher-risk, but if it worked could be a complete paradigm shift. This latter approach from Caltech, led by Professor John Bercaw, involved looking at a reaction invented in Russia some thirty years earlier, the Shilov reaction, which used a platinum catalyst, and making it work to convert methane to methanol. All the faculty, both the superstars and the younger faculty, were inspirational. So while we could not find any reason to prefer one or the other, I felt sure I could get them to work together, and to do that we agreed within BP to give each $10 million. Illinois had been good but not at the same standard as Caltech and Berkeley, and Texas A and M, while having some great scientists, didn't involve the engineers in their proposal at all. Indeed, I was getting calls on the weekends from people who had been excluded from discussions trying to undermine their colleagues. Ah well, academics!

The climate change centre was also interesting. While the problem posed was how to deal with CO_2 emissions, and in particular how to explore whether carbon capture and storage was a viable approach, the three universities, MIT, Stanford and Princeton, each saw this problem completely differently.

For Stanford, if CO_2 was going to be stored deep below the surface, this was a problem of understanding the subsurface geology completely. Stanford proposed both theoretical and computational programs for advancing this understanding.

At MIT, they thought that this was really an engineering problem. We have the CO_2, mixed with other gases, 'here', and we have to get it separated and move it to 'there', and then we have to make sure it stays there for a long, long time. So, they proposed, let's break this down and engineer all the key steps to perfection, watching along the way to see if we can do it cheaply enough so that it is feasible.

The Princeton gang said that, well, there are certainly some geological issues, and we have someone to work on those, and there are some applied science and engineering problems, and we will work on those, but at its heart this is an environmental problem. We are taking a gas which is viewed now as a pollutant and going to store it someplace. MIT was proposing to store it at the bottom of the ocean. Will the public not say, 'OK, so you guys screwed up the air, and now you are going to pollute the ocean too?', or if it is to go underground, will you not have to deal with issues of how long it will stay there, what happens if it escapes, how does the storage perturb the geology, and a range of other issues?

After listening to all this, we were on the plane back to the UK, and my colleague Paul Rutter came over to my seat to chat. In the lounge at the airport we had done a straw poll of the three of us, and each university had one vote! But it was Paul who most clearly artic-ulated the three different approaches that the universities had taken, and talked me into believing that the Princeton approach was the correct one. He was right in pushing the decision in that direction, and the work that took place under our sponsorship, while not advancing CCS very much, did influence how much of the world thought about solutions to the climate change challenge, thanks to some innovative thinking by Rob Socolow. As important, the Princeton faculty, with lots of innovative and radical thinkers such as Peter Singer and Michael Oppenheimer, influenced the corpora-tion through their contact with us at various visits and meetings.

We agreed to give Princeton $15 million over ten years (and BP has since committed money to continue the programme, now in its eighteenth year). After we made our decision Rob Socolow and Steve Pacala asked how we would feel about them approaching Ford Motor Company to also be a sponsor with us. Of course, that was great from our point of view, because it led to closer cooperation with senior Ford executives across a range of things unrelated to the Princeton programme. Business is about relationships, but there are probably only a few times when scientific cooperation can lead to business cooperation that is more or less unrelated to what we were doing on the science side. This was such an instance.

So what had I learned from the work in setting up these centres? That there were faculty, not all of them to be sure, who had built their careers doing basic research, but were now finding the possibility of doing something that could be of commercial value to be very attractive. These faculty did not lack for research support – we could not have tempted them just by money, though the money we were offering was substantial. But once again, particularly in chemistry, the action had shifted from new reactions and structural tools to modifying existing reactions so as to make a material impact on a very big business.

While I was working on getting these centres going, Andrew Mackenzie came to talk to me. He had been having discussions with John Browne, and they would like to suggest that a new position be created for me. How would I feel about being Chief Scientist of BP? My first thought was that not long before I had assumed I had done my last job for the company and was on my way out, and that Doug Ford, who was my boss during the clean fuels Thirty Cities programme, was very keen to see my exit happen. Now he was gone and I was still here. My second thought was: well, into every career there should come one job that is so special, so good, that you just feel joy at the prospect of having it. Mostly that never happens. This was that moment for me. And for the next several years, until I retired from BP at the end of 2003, I was Chief Scientist.

No one had had this job before, so I could define what the role comprised. Part of it was to do what I had been doing, with the university centres. But now I could put this in a context. Over a few months I decided that there were three parts to the job.

First, provide advice when asked, and sometimes when not asked, to the senior executives of the company on technical matters, with the support and help of outside advisors. BP had long had a

Technology Advisory Council, and I was to be one of two internal members of that council, working with the chairman, Sir Robin Nicholson, a BP independent director, to select members and set the agenda. The council was an important conduit for advice into some of the senior leadership of the company.

Second, to lead the technology input to BP's strategy. There was a Chief Economist and he and I could input to strategy from our two perspectives. Sometimes we agreed, sometimes not, but both of us were respected within the company for our views. And occasionally the Chief Economist and I could decide that an issue needed some study between us, and commission the necessary work so we could bring a strategic issue to the attention of senior management.

Third, and this was the major time commitment, to figure out what was happening in the world of science, broadly defined, that could affect the company on any time scale, near term to fifteen years hence, determine what we should do about it, and then do it. The university centres fell into this category, but they were not always the answer. Nonetheless, I was not done with that approach.

And with all of this, I needed to play a role as part of the technology leadership of the corporation, working closely with Andrew and later with his successor, Tony Meggs.

28
Ni
Nickel

Making Bugs Do Chemistry

O NE AREA OF SCIENCE PARTICULARLY CAUGHT MY attention, and that was the role that advanced biological techniques could play in fuel chemistry and in the chemical industry. One biofuel was already in widespread use, and that was ethanol from corn or sugar beets, all of it made by very old processes. But what if we could get some sort of microorganisms to synthesise a more optimum fuel for us? Certainly the technology seemed to be there in 2000, and was being used in other industries, but on a much smaller scale than would be necessary for fuels.

I thought that leading academic biologists would be interested in this, and spoke to friends at MIT, where there was an outstanding group of molecular biologists. Back came the answer that they were completely committed to the pharmaceutical industry, and really had no interest in working with an oil company. Five years later they changed their minds, when my successor as Chief Scientist looked to establish a major university centre in biofuels.

I was not prepared to drop this, but was unsure how to proceed when the answer called me up. I was contacted by Tom Connolly and John Rainieri from DuPont, who asked if they could come by for a chat in London. Tom was DuPont's research director, and John led their biomaterials programme. Now back when I was working on crystallisation of polyesters there was a series of polymers we studied where the linkage between the 'ester groups' was two, three or four carbons, as I mentioned previously in one system of nomenclature these were written as 2GT, 3GT and 4GT, and more commonly were known as PET, PPT and PBT, as the linkages were ethylene, propylene and butylene. Now it turns out that in making chemicals from oil, it is relatively easy to make even-numbered hydrocarbons (so E and B) but not odd-numbered, P in this case. But the PPT polymer had interesting properties, particularly as a fibre. I had obtained small samples of it from my friends at Celanese back then, but it was not in large-scale commercial production. What DuPont, under Connolly and Rainieri, had done was make major modifications to the genetic system of *E. coli* so that they became little factories for producing the three carbon building block for PPT, propylene glycol. It had taken several years of work with a large team, but this eventually became the commercial fibre product *Sorona*.

Now with this toolbox, Tom and John were beginning to think about whether it would be possible to make a component for fuels via the biological route, and I was thinking the same thing but had no toolbox. I didn't even have a 'mechanic', because BP did not employ a single biologist in its research operation. Their visit to me took place in early 2001, which was an interesting quirk of timing. Some years earlier DuPont had acquired the oil company Conoco, and the natural place for them to turn when thinking about fuels would have been there. However, in 1999 DuPont divested Conoco, so they had to look elsewhere. I never knew whether they asked Shell and Exxon before they asked us, but I didn't care. We were receptive.

After some discussion, we signed a simple confidentiality agreement between us, and agreed that a good way forwards was to have a brainstorming session where a number of our key fuels people, and their commercial/technical experts, tried to figure out what molecule(s) would make sense. I knew the people I wanted in BP (which by that time was BP Amoco), and in due course we gathered for a day at the BP offices in New York. This was an exciting day, the discussion flowed from the beginning and we soon had a board covered with ideas. I realised that when business-oriented scientists from similar very big companies got together, we could quickly and productively interact, in a way that was much more like hard work for business–university or big company–small company interactions. In electrical engineering there is a concept of impedance matching, minimising the resistance to the flow of energy through clever design, and the BP–DuPont interaction was well matched from the beginning.

It took a fair amount of work by the lawyers for our two companies to put the legal framework in place. While both sets of lawyers worked hard to protect the interests of their company, they were also sensitive to the fact that the executives wanted this to happen, so their mission was to make a deal not block a deal. Clear communication of intent between the executive and the legal team is very important in companies.

Today, the discussion that started between us is the commercial venture *Butamax*, a DuPont–BP joint venture which is seventeen years old, making the molecule isobutylalcohol by a biological route. We had shifted chemistry from the older petrochemical syntheses to a living biological reactor.

29
Cu
Copper

Do You Like Mushrooms? Not Just in the US and Europe, but Also in China

WITH CAMBRIDGE GOING WELL, CALTECH–BERKELEY and Princeton established, John Browne asked me to turn my attention to China. BP was trying to develop on several fronts in China, and he suggested that we try to establish a university research programme there similar to what we had done at Cambridge and in the US. Now I knew quite a lot about US and UK universities, and so that had been easy, but China – where to even start? I began by asking friends at various US and UK universities who they thought were the leading thinkers in energy in China, and was quickly pointed to Tsinghua University in Beijing, which I already knew to be one of the leading places, but now I had names of some of the key people doing research in energy, especially on the policy end.

I also got some negative comments, mainly from people who did not know China well, but who had their prejudices. One well-known professor at a UK university said, 'Well, you won't find a Cambridge in China you know,' implying that if we were going to fund research there it would be second-rate compared to what we had done in the UK and the US.

Of course, I had to coordinate all of this with the leadership of BP in China. Fortunately, the President of BP China was Gary Dirks, a good friend who had actually been at the Sohio lab in Cleveland with me many years earlier, come to London, then to Brussels a while after I crossed the Atlantic. So he had a technical background, and saw the value to the company in our setting up something like a research centre in China.

I had a good relationship in the UK with the Royal Society, in part because we gave them a substantial chunk of money each year for certain programmes, to support both women in science and international cooperation. One day, while I was musing about how to tackle the China centre, I was invited to a lunch at the Royal Society being held in honour of the President of the Chinese Academy of Sciences, who was visiting.

Now in Russia, the Academy of Sciences is not just an honorary organisation, but runs its own laboratories and degree-granting institutions, in parallel with the universities. The Russians had heavily influenced the development of the Chinese scientific system, especially in the years from the revolution until the later part of the Cultural Revolution. China thus also had a large number of laboratories run by the Chinese Academy of Sciences (CAS), and a little research on my part showed that several of these were quite prestigious.

I took advantage of the luncheon to meet the President, and to tell him that I wanted his help in starting a research centre in China dealing with energy matters, both laboratory science and policy. He asked that I write to him outlining my thoughts, and the size of our potential commitment, and how I would like to go about it. I did this, and at the same time I told him how we had done the US centres, including the idea of competition between laboratories, though I said that perhaps that was not a suitable approach for China.

When the President of the CAS replied, he suggested several possible laboratories which we might consider, especially commending to me the Dalian Institute of Chemical Physics. I had told him that I might want to also look at universities, and in particular Tsinghua University in Beijing, and the University of Science and Technology of China (which I later found out was run by CAS), and he was very amenable to that as well. But of interest to me was that he said that of course we should have the research groups compete for our funding, and he would very much like to see which ones we felt were the best. This was to prove a theme of my work in China, a passion for excellence and a desire to have their own views of quality validated by assessment from outside China. He offered to assign one of his staff to me, to make suggestions of places for me to look at, and to help us in any way we needed. Well, I thought, this could be fun.

I arrived in Beijing one day in 2000, and was met at the airport by BP drivers. BP insisted that all cars I used be provided by them, rather than by CAS, so as to ensure that our standards of safety, including seat belts being worn and drivers not talking on the phone while driving, were enforced. At the BP office I met with Gary Dirks and the BP staff, and reviewed the schedule and learned their aims for what I was doing, which were mainly to be sure that we conducted ourselves so as to have favourable publicity and substantive benefits in terms of government relations from what I was going to do. Then I met with Mr Wang, the young man assigned to me by the CAS president.

Our first site visit was to be Shenyang, a very old city in northern China, where they felt interesting work was being done. After that I was to visit the Dalian Institute of Chemical Physics, in northeastern China, not far from the North Korean border; a research laboratory specialising in lubricants in Lanzhou, which is thought of as being in western China, although there is a lot of China west of Lanzhou; the University of Science and Technology of China, in Anhui Province to the south, where some of the most advanced facilities, such as a synchrotron, were based; a laboratory specialising in oil and chemicals related research in Taiyuan; and finally back to Beijing and a visit to Tsinghua University, where there was research across a range of subjects, but I was particularly interested in their work on energy strategy for China.

We landed in Shenyang in the early evening, and it was arranged that we would meet with senior faculty for dinner. As we left the hotel, Mr Wang asked me a strange question: 'Do you like mushrooms?' Well, of course, mushrooms, sure. It turned out the restaurant that we went to brought a big bowl of steaming broth to the table, and then for the next hour or so a selection of different mushrooms, maybe twenty or thirty different ones, that we cooked and ate. Only mushrooms. Ah well, this was my first but not my most unusual meal in China.

Food, and eating together, is very important in China, and almost every meal I have ever had there has been communal dishes, and lots of them. On the other hand, meals are quick, not the three- or four-hour French lunch, and while there is alcohol served in the evening, it was far less than the quantities I was later to encounter in Russia at lunch. In Dalian the staff would say, 'We are going to have lunch at fast food.' This was a staff restaurant, where we always had a private room because we were usually accompanied by the director, and where, as we sat down, the dishes would immediately appear on the large turntable in the centre of our table. Twenty minutes later we could be walking back to the conference room to continue our discussions!

In Shenyang I saw some good science, but it was all in the lab of one younger faculty member who had been lured back from the US, and there was not the broad-based excellence I was seeking. But in Dalian it was another story. Here, one scientist after another showed work that was at the frontier of applied chemistry, and the equipment in the labs was also as good, or better, than any lab in the US and Europe. Much of the work was very applied, focused on commercial applications, although the scientists published lots of papers as well. The director, Xinhe Bao, had spent time at some of the best institutions in Germany, was fluent in Chinese, English and German, and, although newly appointed, was very much a leader of the enterprise.

While I was very impressed by a lot of the applied chemistry of methane and related reactions that the Dalian guys were carrying out, in the course of discussions Director Bao mentioned that there was also a lab on campus doing work on fuel cells. I asked to see this, and in due course was taken to a part of the campus I had not yet visited, and saw some of the products they had produced, including a small power pack for bicycles. (At that time there were still lots of bicycles on the streets in China; indeed in my first visit to Beijing there was one lane each way for cars down the main road past Tiananmen Square, and all the rest of the large boulevard was filled with bicycles. Two years later the bicycles had been crowded into a single lane on each side of the road.) There were lots of other interesting fuel-cell products as well, of a design and originality I had not seen elsewhere. But when, during a conversation the next morning, I expressed an interest in supporting this work, Xinhe Bao told me that would not be possible. Most of the fuel-cell work was for military applications, and they would not open any of it up for external support. Indeed, they had only shown me some of their finished products, and none of the laboratories where they developed them, nor did I get to speak with the scientists and engineers involved in the work.

The University of Science and Technology of China, in Hefei, Anhui Province, started out in Beijing, but many years ago, when the Chinese feared a Russian invasion, they moved it to someplace considered more secure, that is, far away from everything else. Nonetheless, it is not more isolated than many American Midwestern universities. The labs were very well equipped, there were major research facilities there, and the university president turned out to be very switched on. He had travelled around both the US and Europe, visiting many of the technological universities considered to be excellent on each continent, and tried to use that experience to shape the development of his campus. One of the interesting things they do there is to search out mathematical prodigies from across China. These come to USTC around age eleven, and a parent must come and live with them. By age fifteen they are considered ready to get a PhD, and are sent off to the US or UK for this. With 1.3 billion people to choose from, they manage to keep the special dormitory full.

I had a good day listening to science there, but again I did not find enough of it compelling in terms of what we wanted to fund. Still, over dinner there were interesting discussions with the president and his colleagues as course after course was served to us. At one point they stopped putting dishes on the communal turntable, and a waiter stood behind each person with a plate covered in a metal dome. These were placed before us and with one swoop the domes were removed. When I looked at my plate, there was a whole turtle, shell, head sticking out, tail, four little feet. It had been cooked and some sauce was drizzled over the shell. I turned to Mr Wang and said, 'Well we have had many meals together in the course of this trip, but you have now presented me with something that I don't know how to eat, and I am pretty sure I don't want to eat.' No, no, they all protested, at the very least take the shell off with your chopsticks and taste the sauce on it, very delicious. I did this, but realised that they should have added, don't look down, because when I did there you could see the entire digestive tract and organ system of the cooked turtle. The sauce was tasty.

I travelled out to Lanzhou with Mr Wang and one of the elder statesmen from Dalian, as the Lanzhou institute had originally started as a branch of Dalian. We had a long journey, including a stop in Xian. During the flight I asked him whether, after our trip, he was returning to Dalian. No, he said, his mother had died two years earlier in Shanghai and there was a family gathering to commemorate this. Gradually he told me his story. He grew up in Shanghai when it was divided into various western concessions, living in the German area. As a boy, despite the Japanese invasion, he was able to get an education, and after the war lived through the turmoil of the ascent of Mao. He completed university and was sent to Russia to study for a PhD in chemistry. There he met his wife, also a Chinese PhD chemistry student. After they returned, they were both employed on the faculty of what was then the oil and gas technical institute in Dalian, later to become the Dalian Institute of Chemical Physics. Dalian had actually been occupied by the Russians, both in the early part of the twentieth century, and then again after the Second World War, the last Russian troops only withdrawing in 1955. Under Mao there was already suspicion of those who had been educated in Russia as relations between China and Russia worsened. Nonetheless, they built up the institute, getting some equipment and doing good science. When the Cultural Revolution came, the professor and his wife were both sent to the countryside to work on farms and for political re-education, while the director of the institute was, for a time, locked in a cage in the basement of the research building. When they eventually were allowed to return, everything had been wrecked, all windows smashed, the laboratories in shambles. So they gradually rebuilt until they could do research again. Then, he said, along came Deng Xiaoping, and asked whether they could do any research that might lead to development of businesses for China. Oh, OK, they said, if you want practical businesses we can do that too. What a scientific life, I thought; in comparison, mine had been rather tranquil and ordinary.

Finally, after eating noodles and drinking sorghum liquor in Lanzhou, while hearing about lubricants research, we made it back to Beijing, first to visit several staff at Tsinghua University, and then for high-level discussions with the leadership of the Academy of Sciences, which our country president, Gary Dirks, also attended. Tsinghua turned out to be great. Rob Socolow at Princeton had recommended to me that I meet Professor Ni Weidou there, and I spent a fascinating day with him and his younger faculty talking about energy issues, and how China might respond. It was the first of many such discussions.

The Chinese scientists I met then and later were all very much abreast of the latest developments in their fields, and the leaders of the various laboratories and universities, and those centrally in Beijing, were passionate about excellence. They wanted to know how good their labs were, and to see if our opinions validated their own assessments. They had developed a system for ranking their laboratories, and were rigorous about directing more funding to the best ones, and looking at improving or closing the worst. If only every country took such an approach!

Over the course of several visits, in which I once again involved other colleagues from BP, we settled on support of Dalian for work on methane that would complement that at Caltech and Berkeley, and work on energy strategy and climate change at Tsinghua that would go well with the work being supported at Princeton, especially as the Princeton and Tsinghua groups already knew one another. Taking a bit of a leap that the Chinese work was as good as I thought it was, I invited Xinhe Bao to come and give a talk at the BP (formerly Amoco) research centre outside of Chicago. I could see when the local staff filed into the auditorium it was with a mixture of curiosity and disdain. Then Bao began to present the work of the lab, and soon they were leaning forwards, and then asking questions, and, an hour and a half later, left completely impressed by what they had heard. That gave me confidence to put him together with Enrique and Alex from Berkeley, and they accepted the Dalian guys as scientific peers. So, no, as one of my UK commentators had said, I did not find Cambridge in China – no lawns, gowns and centuries of professorships – but I did find science and engineering comparable to any being done elsewhere. And when I reflect on this now, I think, what a pity that these great, inventive, scientific minds, because of political upheaval, missed out on being involved in the solution of many of the great chemical problems of the second half of the twentieth century.

When Gary Dirks and I met with the leadership of the Academy of Sciences, perhaps a second meeting after we had clearly formulated what we were going to support, the president of the Academy turned to me and said, 'You know, this is interesting that you have come here three times, visited all these laboratories and formulated a programme of work to support. Because one of your competitors among the large oil companies came here six months earlier than you, and asked to sign an agreement with us to cooperate on science, held a press conference, and we have never seen them again. I like your approach better.' Well, that is a nice lesson, and consistent with one that John Browne always taught us, not to open up a rhetoric gap. Less talking about what you might do and more actual doing is a much better use of everyone's time.

The Chinese named our joint research centre 'Clean Energy – Facing the Future', and when John Browne and I had a chance to meet Hu Jintao on a trip he made to the UK shortly before assuming the leadership of the country, he said our centre was an act of 'strategic vision'. It also led to some very good applied science and engineering.

Before I left BP in late 2003, to start my third career (after my academic life and my big company industrial life), I was invited to speak to a large group of leaders discussing energy policy and technology in the Great Hall of the People in Beijing. What I said then was not completely correct, viewed from fifteen years of hindsight, but not completely wrong either. It bears on what I have been saying about the future of energy, of university research and industrial development, and about excellence. After reviewing some of the developments in energy, I turned to the question of supporting basic science, because I felt then, and feel now, that there is always a great danger that governments think they can only do technology, and applications, without doing any basic science. I said, in part:

> What is the science behind our energy needs? Well, there is a lot happening in science today but the great excitement is certainly around information technology, biotechnology and materials science, including

nanotechnology. I am convinced that all of these will influence the big possibilities for energy for the coming decades, and that it is absolutely essential for any country that aspires to leadership in energy technology to have a strong basic science effort behind the technology. Building on the basic science is applied science, and building on that is technology development. This is a chain, and it is a mistake for a great country to not have all three links in the chain.

How does a country justify its investment in basic science? After all, some will argue that since the results of basic science are available to everyone through publication, it can be left to other, richer countries. I reject this idea. Basic scientific research attracts great minds, great intellects, and their intellectual vitality is important for any country. They train students, most of whom go on to more applied careers, but with an understanding that can never be gained if you only enter at the technology end of the chain.

Shortly thereafter I retired from BP – but not from being active in business, in science, in technical developments. But my life of active involvement in science, even to the extent I had done it in my chief scientist role, largely ended and transitioned to a role of venture capitalist, board member, government advisor and lots of other things, all interesting but all part of another story than the one I have told here. I used my scientific brain, and continue to use it every day, and certainly have made use of what I think of as my most important core competency, namely the ability to learn new things, but just as I have asserted that chemistry was moving into a new phase, so was I.

Celebration of Mastery for the Discipline of Chemistry

I WANT TO CELEBRATE THE DISCIPLINE OF CHEMISTRY and its great achievements. Chemistry remains a wonderful field for students to pursue, because it is so central to solving problems across almost all aspects of our lives. I hope I have demonstrated that a student who is successful in chemistry at the undergraduate or graduate level will have acquired a way of thinking and a way of learning that will serve her well no matter what she chooses to do. But recognising this does not in any way change my view that the big research problems fundamental to chemistry have been solved.

What remained to be done in 1950? There were lots of questions about molecular structure that still needed solving. Most of what had been accomplished to that point was from crystal structures of organic compounds using X-ray diffraction. Both the three-dimensional arrangements of atomic nuclei in molecules, especially inorganic and organometallic compounds, and the details of these arrangements, such as bond lengths and bond angles, had to be determined. The symmetry of molecules provided some of the information about three-dimensional arrangements using infrared and Raman spectroscopy, nuclear magnetic resonance provided additional information, especially as it extended beyond just the environments of hydrogen to fluorine, carbon and other nuclei.

Regarding reactions, before 1950 products of reactions were often identified by carrying out other reactions. As undergraduates, we took courses in qualitative analysis and qualitative organic chemistry, in which we learned tests involving colour changes for specific ions

---

CHEMISTRY RESEARCH TODAY

I looked at the weekly news magazine of the American Chemical Society, *Chemical and Engineering News*, to see if it reflected this development. This magazine always has a mixture of science, engineering and business news related to the chemical enterprise, so I turned to the pages of scientific developments that they reported. I took five issues from the spring/early summer of 2016, and then I went into the archives online and randomly picked some issues from 1959, trying to use the same standard to determine what was basic chemistry being reported and what was application of chemistry to other fields, such as biology, materials science, environmental or agricultural problems. OK, I know that this is not 'scientific' or statistically significant, it is just an indication, a feel, for what is going on and what was going on.

In the issues I looked at from 1959, there were twenty-nine articles about scientific developments in chemistry. Of these, twenty-two were basic chemistry according to my classification, and seven were application of chemistry to other fields, i.e. about 75 per cent basic and 25 per cent applications. The basic chemistry covered a very wide range, including both organic and inorganic reaction mechanisms, structure and bonding in a class of molecules called metallocenes (discovered in

or molecular species when a particular reagent was added. Then we carried out additional reactions to make solid compounds and measured their melting point, hoping to determine, in this way, their identity. There were many volumes of such compound melting points used by chemists around the world. This approach limited the throughput of chemistry. A five-minute infrared spectrum could determine if a reaction had yielded an aldehyde, a ketone, an alcohol or all three. Separation techniques such as gas chromatography allowed for each compound to be recovered, and when combined

---

1951 and whose structure, that of a sandwich of two rings on either side of a metal atom, was elucidated definitively in the mid-1950s), a number of analytical techniques including chromatography, thermal analysis and infrared intensity measurements. There were also several reports of new polymers. The seven application articles were about chemicals found in space, chemicals derived from insects, carcinogens in cigarette smoke, a new antibiotic, a way to block the development of ulcers and enzyme chemistry. In other words, they were heavily weighted towards biomedical applications. Interestingly, in light of what I am about to tell you about 2016, catalysis did not figure at all in these articles.

The issues I looked at from 2016 contained fifty-five scientific briefs. Of these, I found, forty-one were applications, six were catalysis and eight were basic chemistry. If we count the catalysis as being basic chemistry, and I think that is correct, then the split is approximately 80 per cent applied and 20 per cent basic. The basic articles were about a new spectroscopic method, a few novel synthetic routes and some aspects of chemical bonding. The applications were heavily weighted to biology/medicine, comprising more than half of the applied articles, then materials, environmental and one each on food, art, cement chemistry and nuclear weapons. So while this is a bit random, it is, I think, a good reflection of the change that has taken place.

with mass spectrometry the detailed connection of atoms in the molecule to be determined unambiguously. Chemistry had to correct mistakes from the past, and practically invent inorganic chemistry as a science rather than a huge collection of observations. In parallel to this work, it was possible to see patterns and form an understanding of how broad classes of reactions occurred. For example, at University College London during the period 1930–70 Sir Christopher Ingold was able to show that a very large number of organic reactions, thousands of them, fell into one of two groups in terms of mechanism.

The theory of structure and bonding known as Valence Bond worked pretty well for organic molecules, even if it required some fudging when it came to aromatic structures such as benzene, naphthalene, etc. and their derivatives. But for most of inorganic and organometallic chemistry it was hopeless. Once molecular orbital theory, founded in quantum mechanics, was brought to bear, the broad range of experimental results, new compounds such as metallocenes, and many other observations became part of a coherent explanation that had heretofore been lacking. And as computer power increased, and became available to chemists, many properties could be calculated and compared to experimental results.

Chemistry developed a model if not a rigorous theoretical understanding of how reactions occurred in the gas phase fairly early in the twentieth century, and how atoms/molecules behaved as gases. But the complexity of condensed matter – liquids, amorphous and crystalline solids, liquid crystals, synthetic and naturally occurring polymers – required the full complement of theory and experiment that could be brought to bear in the second half of the century. To understand phenomena like crystallinity even in the simplest polymer, polyethylene (a long chain of CH_2 groups), was a monumental effort involving many of the leading minds during the 1970s and 1980s.

And reactions of importance to industry often occur through what is known as heterogeneous catalysis. Various species coated onto a flat solid surface, or onto some sort of inert beads, speed up a reaction in the gas phase or in a liquid phase in contact with the surface (so heterogeneous because they occur in at least two different phases). Again, a huge effort, which continues today, was required, with many new techniques brought to bear, to understand how these reactions occurred, what made them selective for a particular desirable product and how to improve their efficiency. This is largely what I referred to in the opening of this book as putting the chemical industry on a more scientific basis.

I know there will be those who disagree with my assertion that the major problems of chemistry were largely solved in the second half of the twentieth century, that what remains is filling in a few places that were not quite finished, that the big work is done. I think that many who take that view will be devoting their lives to completing that work, and no one wants it to be implied that their work is not just as important or impactful as that which went before. I am not suggesting that. But I do feel that there is a very much reduced scope for new insights in this discipline.

And it is true that in this final phase there have been a couple of big discoveries that shook things up – perhaps the biggest being the new forms of carbon – the fullerenes and nanotubes, as well as graphene and its properties. Fullerenes were first positively identified in 1985, but they were predicted and modelled twenty years earlier (though many did not take this prediction seriously), so were very much a part of the chemical discovery enterprise of the second half of the twentieth century. Graphene was likewise understood as a possibility for several decades before the simple way of producing it was discovered, and even many of its unusual properties had been predicted. These discoveries are worth mentioning because the challenge to my thesis of chemistry being pretty much wrapped up is that there could still be some big surprises out there. I don't think so. Likewise, the last big technique development was the scanning tunnelling microscope, developed in 1981. While there have been major advances in all the techniques for studying structure and bonding, and these continue, they are evolutionary, whereas the STM (and the related Atomic Force Microscope) was a step change in our ability to image at the atomic level. And again, it occurred squarely in the second half of the twentieth century. Perhaps the last big development on reaction chemistry is metathesis, which Yves Chauvin described in 1971, including what sort of metals would catalyse the reactions, though the catalysts that were eventually developed and put to use were made by Richard Schrock and Robert Grubbs in 1990–92, with the three of them sharing the 2005 Nobel Prize. So there have not been any big, unexpected discoveries in chemistry – materials, structures, reactions, techniques – for more than thirty years, or, if we date from Bob Grubbs' paper, twenty-five years.

The accomplishment of the last half century and more is monumental. We understand structure, bonding, reactivity, the nature of solids, liquids, liquid crystals, polymers, both simple and complex, at a level of detail, both from experiment and theory, that no one could have envisioned was possible in 1945. This from a combination, as I have asserted, of a range of spectroscopic, mass spectroscopic (which is not spectroscopy at all as I have introduced it, but rather a series of techniques for blasting molecules with high energy and elucidating their structure from the fragments that result), chromatographic (that is separations technology) and other analytical techniques, of which I have described only a few (some of them made possible by the invention and commercialisation of a wide variety of lasers and imaging technology), and the massive power of computers to acquire and process data as well as to do calculations routinely, in seconds, that were a PhD thesis as recently as thirty years ago. To all this hardware has been added huge developments in the theory of structure, bonding and reactivity.

And what of the chemists doing research in the leading chemistry departments around the world? Are they obsolete? No, certainly not. For the most part, they are using their exceptional scientific insights and determination to solve very complex problems in catalysis, materials, biology and medicine, as well as tackling environmental issues of great importance. It really wouldn't matter if they were doing it in departments that universities named something other than chemistry, the work would be just as valid and just as interesting. Someone once said, 'Industry has problems, universities have departments', but at the very best places over the past few decades universities have allowed the boundaries of what is done in research in those departments to be very fluid. If you visit a top-rated chemical engineering department you will find research being done that could be in an equally highly ranked physics department, and in many chemistry departments the collaborations and joint appointments extend to medicine, computer science, materials science and biology.

Meanwhile, the educational mission of these chemists, just as for the chemical engineers, remains a crucial one at both undergraduate and graduate levels. As I have tried to stress, taking an intelligent high-school graduate and turning her into a chemist is more than teaching her the nomenclature or reactions of organic chemistry, or how to solve certain well-known problems, or the theory of bonding, or even thermodynamics and quantum mechanics.

Indeed, it is about more than learning how to measure and observe, important though those skills are. In the end, it is about taking a still-plastic brain and teaching it the unnatural way of thinking that we call scientific, and this is what the faculty do in first-rate undergraduate and graduate programmes. And for this, it doesn't really matter if the research they are doing is pushing the boundaries of fundamental chemical science or, more than likely, advancing biology, medicine or nanotechnology.

Chemistry has given me, and thousands like me, a rich and wonderful life. The people that I studied with as an undergraduate, a graduate student, and those I worked with in several jobs, as well as the great communities of scientists of which I was a part – we had the privilege of being able to be a part of painting the beautiful masterpiece that is today's chemical understanding, and being in our own ways small contributors to laying the foundations that enable the advances in so many diverse fields today. And for that we thank our parents, teachers and colleagues, the governments and charities that supported us, the journals and publishers that disseminated our work and everyone who considered it important that this science be done.

Acknowledgements

THE NOTION OF THE COMPLETION OF THE chemical discipline first arose during a series of conversations I had with the late Professor Arnost Reiser during one of his regular visits to London. It was reinforced by rereading an article by Professor Peter Atkins in John Brockman, ed., *The Next Fifty Years* (London: Weidenfeld and Nicholson, 2002), in which he said, looking forward to the first half of the twenty-first century, that 'at the beginning of the twenty-first century chemists are in complete command of matter'. He also asserted, as do I, that 'It must be admitted, however, that little new understanding of chemistry itself will emerge over the next fifty years. The subject is already highly mature, and there are unlikely to be many surprises in connection with its fundamental principles ... we can be confident that all such discoveries will fall within our current canon of understanding.' A similar view was taken by Professors John Deutch and George Whitesides in a 2011 editorial in *Nature*, in which they urged a redirection of the academic chemical enterprise to the solution of problems in adjacent disciplines.

I owe my scientific life to many teachers, colleagues and friends, but especially to Professors Robert Bauman, Walter Edgell and Hans Gunthard, to my many graduate students and postdoctoral fellows, and my collaborators Eli Pearce, Menachem Lewin and Iain Dea. Jeanette Grasselli and Glenn Brown took a chance hiring me into Standard Oil more than thirty years ago and gave me the opportunity to become an industrial scientist. Later Andrew Mackenzie, Rolf Stomberg, Rodney Chase and John Browne put me into positions where I could influence not just a very big corporation but an entire industry. I am grateful to all of them.

About the Author

BORN IN A NEW JERSEY FARMHOUSE, EDUCATED on the streets of New York and at several fine universities, Bernie Bulkin is by now a fully paid up member of the British establishment and in the midst of his third career. He spent 18 years as an academic scientist, teacher, and leader at City University of New York and Polytechnic Institute of New York, publishing about 100 articles and two books. For his research he received the Coblentz Award, the Society of Applied Spectroscopy Gold Medal, a Sigma Xi Distinguished Research Citation, and the Oscar Foster Award in Chemical Education. He held positions of Department Chair, Dean of Arts and Sciences and Vice President for Research and Graduate Affairs. He also served on the Boards of Optronics International and Spex Group during this time.

In his second career, also 18 years long, he held a variety of industrial management and research positions with Standard Oil and BP, including Director of Analytical and Environmental Sciences, Head of the Products Division, Vice President for

Refining, Chief Technology Officer for BP Oil, Vice President Environmental Affairs, and Chief Scientist.

After leaving BP in January 2004, Bernie Bulkin has been a venture capitalist with California firm Vantage Point and London firm Ludgate Investments Ltd, he has been on the board of 11 companies, chairing two UK public companies, and has held several posts with the UK Government, including Chair of the Office of Renewable Energy. His current board memberships include ATN International and ARQ Ltd. Bernie has held numerous positions with educational and charitable organisations. He is Emeritus Professorial Fellow of the University of Cambridge and a Vice President of the Energy Institute. His radio programmes, *Environment on the Edge*, were heard on Voice America, he has contributed regularly to Huffington Post, and he is the author of the 2015 book on leadership, *Crash Course*. Dr Bulkin is a Fellow of the Royal Society of Chemistry and a Fellow of the Energy Institute. He was made an Officer of the Order of the British Empire (OBE) by the Queen in the New Year Honours list 2017.